GeoTourism for Social Sustainability and Ecological Justice in Sub-Saharan Africa

Unlocking the Potential

Edited by Khodani Matshusa and Llewellyn Leonard

Routledge
Taylor & Francis Group
LONDON AND NEW YORK

First published 2025
by Routledge
4 Park Square, Milton Park, Abingdon, Oxon OX14 4RN

and by Routledge
605 Third Avenue, New York, NY 10158

Routledge is an imprint of the Taylor & Francis Group, an informa business

British Library Cataloguing-in-Publication Data
A catalogue record for this book is available from the British Library

Library of Congress Cataloging-in-Publication Data
Names: Matshusa, Khodani, editor. | Leonard, Llewellyn, editor.
Title: Geotourism for social sustainability and ecological justice in Sub-Saharan Africa : unlocking the potential / Edited by Khodani Matshusa and Llewellyn Leonard.
Description: Abingdon, Oxon ; New York, NY : Routledge, 2025. | Includes bibliographical references and index.
Identifiers: LCCN 2024040750 (print) | LCCN 2024040751 (ebook) | ISBN 9781032706733 (hardback) | ISBN 9781032706740 (paperback) | ISBN 9781032706764 (ebook)
Subjects: LCSH: Geotourism—Africa, Sub-Saharan. | Sustainable tourism—Africa, Sub-Saharan. | Sustainable development—Social aspects—Africa, Sub-Saharan. | Urban ecology (Sociology)—Africa, Sub-Saharan.
Classification: LCC G155.A357 G46 2025 (print) | LCC G155.A357 (ebook) | DDC 338.4/79167—dc23/eng/20240925
LC record available at https://lccn.loc.gov/2024040750
LC ebook record available at https://lccn.loc.gov/2024040751

ISBN: 978-1-032-70673-3 (hbk)
ISBN: 978-1-032-70674-0 (pbk)
ISBN: 978-1-032-70676-4 (ebk)

DOI: 10.4324/9781032706764

Typeset in Times New Roman
by Apex CoVantage, LLC

GeoTourism for Social Sustainability and Ecological Justice in Sub-Saharan Africa

This book provides a comprehensive overview of the potential of geotourism in sub-Saharan Africa for promoting social sustainability and ecological justice. It examines the challenges facing the region for geotourism development and the innovative approaches and best practices for overcoming these.

Drawing on case studies from the region, the book provides innovative insights that explore the intersection of geotourism with poverty alleviation and job creation. Thematically it engages with a range of critical issues, from leveraging geotourism for local socio-economic development to the intersection of geotourism with environmental conservation efforts. The book delves into the intricacies of geoheritage and geotourism threats in the face of climate change and extractive industries. By examining the untapped potential of geoheritage sites, this book seeks to shed light on opportunities for promoting local sustainable development in economic, social, and environmental dimensions.

The book will be of great interest to academic scholars, including undergraduate and postgraduate students, industry, governments, and tourism and conservation organisations. It will be of appeal to individuals in the fields of geotourism, ecotourism, cultural tourism, sustainable tourism development, geoeducation, and geoscience.

Khodani Matshusa is a professional natural scientist (geological science) and a researcher at the Department of Environmental Sciences, UNISA. He holds a PhD in Tourism and Hospitality (University of Johannesburg) and MSc in Earth Sciences-Geology (University of Venda). He is the first black African (and currently the only one) with a PhD in geotourism, geoheritage, and geoparks. His research focuses on geoheritage, geotourism, geoparks, social sustainability, and mine environmental impacts.

Llewellyn Leonard is Distinguished Professor in the Department of Environmental Sciences at the University of South Africa (UNISA). He also serves as the Chair of the Centre of Excellence in Adaptation and Resilience at UNISA. He holds a PhD from King's College, University of London. His research interests encompass a wide range of critical areas, including environmental justice, human rights, urban risks, sustainable tourism, adaptation and resilience, democracy, and governance.

This book is dedicated to the resilient communities of sub-Saharan Africa, whose unwavering spirit and profound connection to the land inspire us all. It honours their enduring quest for social sustainability and ecological justice. May this work serve as a beacon of hope and a catalyst for positive change, unlocking the potential of geotourism to create a more just and sustainable future for all. This book is also dedicated to the African continent, a vital part of our mother Earth that must be cherished, harnessed, and protected.

Contents

Figures

Tables

Contributors

Daniel Kwadwo Asiedu is Professor of Geology and an internationally recognised researcher in the field of sedimentary geology. He has over 25 years' university teaching and research experience. His current research activities are focus on understanding the geology of sedimentary basins of Ghana. He has also served in various administrative positions within the university system. He is a fellow of the Ghana Academy of Arts and Sciences, Geological Society of London, and the Ghana Institution of Geoscientists. Daniel is a strong advocate for revolutionising Geoscience education in Ghana, which was the topic of his inaugural lecture at the University of Ghana in 2019.

Bruce Kofi Banoeng-Yakubo is Professor of Geology and an expert structural geologist and hydrogeologist who has carried out extensive work towards understanding Ghana's geology and hydrogeological systems. He has over 30 years' university teaching and research experience. Apart from service in academia he has also worked as Chief Director, Ministry of Lands and Natural Resources in Ghana. He has mentored and supervised many students to various levels in academia and industry and plays an active role in infrastructure development at the department. He has a keen interest in promoting and harnessing geoscience for sustainable development in Ghana.

Hyacinthe Zouyane Nouhou Dama is a PhD student in Petrology and Geoheritage in the Department of Earth Sciences of the University of Maroua. She obtained her MSc degree in 2019 in the same institution on the topic Petrographic characterisation of the geomorphological features of the Lara locality (Far North Cameroon) and inventory of the geotourism assets. She is co-author of some scientific papers and book chapters. Currently, she is a Junior Expert in Geoscience and Environment/Sig at the Global Spatial Environment and technology consultancy.

Merlin Gountié Dedzo is Associate Professor in the Department of Life and Earth Sciences at the High Teacher Training College of the University of Maroua since 12 December 2011. He holds a PhD under the theme Cartography, petrography, geochemical characterisation and magnetic studies of Miocene ignimbrites of Mounts Bambouto and Bamenda (West Cameroon): conditions of setting and location of their emission centres on March 20, 2012, at the University of Yaoundé 1. He is specialised in Geomagnetism-Petrology-Volcanology and has numerous articles in this field. He has been a lecturer since November 2013 and since then, he supervises Master's theses. Currently, he is the Secretary General of the Cameroon Association of Geomorphologists.

Armand Kagou Dongmo is Professor at the Department of Earth Sciences at the University of Dschang since 2022. He has held positions of responsibility in the administration of the University of Dschang for several years. He holds a PhD from the University of Yaoundé 1 in 2006 on the theme Mount Manengouba: Volcanological evolution, magmatological characteristics and natural hazards; comparison with Mount Bambouto and Mount Bamenda (Cameroon Line). His speciality is Volcanology, Petrology, and Geochemistry. He is the author of numerous articles and supervisor of many master's and doctorate/PhD works.

Felix Ekwabena Donkor is a post-doctoral research fellow at the College of Agriculture and Environmental Sciences, UNISA, South Africa. His research interests include sustainable rural livelihoods, indigenous knowledge systems, environmental governance, and sustainable development. He is a founding member of the Inala Food Sovereignty and Climate Justice Forum at the University of the Witwatersrand where he completed his doctoral studies and co-manages a food garden which is entirely for feeding hungry students in his spare time; and is a budding farmer. His hobbies include cycling, swimming, jogging, writing, and music.

Michael Dyssel is a geographer that lectures in the Department of Geography, Environmental Studies and Tourism at the University of the Western Cape, Bellville, South Africa. His research and teaching interests include urban-ecological challenges, conservation, communities and resource-use challenges, environmental management, and the impacts of tourism. Transdisciplinary and multi-methods research approaches underpin his work.

Llewellyn Leonard is a Distinguished Professor in the Department of Environmental Sciences at the University of South Africa (UNISA). He also serves as the Chair of the Centre of Excellence in Adaptation and Resilience at UNISA. He holds a PhD from King's College, University of London. His research interests encompass a wide range of critical areas, including environmental justice, human rights, urban risks, sustainable tourism, adaptation and resilience, democracy, and governance. His work has been pivotal in addressing contemporary environmental challenges, focusing on the intersection of environmental and social issues. He is interested in how power is distributed and how societies can adjust to environmental changes and enhance their capacity to recover from environmental disruptions.

Yvonne Sena Akosua Loh is a hydrogeologist and Senior Lecturer. She has 15 years of experience in university teaching and research and extensive knowledge and expertise in groundwater resource assessment and management. Her current research focuses on groundwater governance for sustainable irrigation and other uses. She is a member of various professional organisations. She is currently serving as Chair of the mentoring and education committee of Women in Mining Ghana (WiMGH) and a council member of the Ghana Institute of Geoscientists (GhIG). She is passionate about mentoring girls in STEM and creating awareness about the importance of geoscience in solving environmental problems for sustainable development.

Khodani Matshusa is a professional natural scientist (geological science) and a researcher at the Department of Environmental Sciences, UNISA. He holds a PhD in Tourism and Hospitality (University of Johannesburg) and MSc in Earth Sciences-Geology (University of Venda). He is the first black African (and currently the only one) with a PhD in geotourism, geoheritage, and geoparks. He was a United Nations Educational, Scientific and Cultural Organization (UNESCO) expert on geoheritage, geotourism, and geoparks, and was instrumental in development of the South African database of abandoned mines for South Africa's Department of Mineral and Energy Resources (DMRE). His research focuses on geoheritage, geotourism, geoparks, social sustainability, and mine environmental impacts. He is the founder of the Union of African Geoheritage and Geotourism (UAGG).

Tracey T. McKay is a graduate of the University of the Witwatersrand, UNISA, and University of Johannesburg. Her research interests include the geography of education, teaching and learning in higher education, and adventure tourism. She has awards for community engagement (environmental education), teaching – a UJ VC award for Teaching Excellence (2011), a commendation by HELTASA (2012), and two Excellence in Scholarship of Teaching and Learning awards at UNISA (2018). She is one of the founding members of the national online Geography Olympiad, SANGO.

Kevin Mearns is a full Professor in the College of Agriculture and Environmental Sciences, UNISA, South Africa. His research interests include Sustainable tourism, Environmental management, Ecology and Ecotourism. He is involved in a number of community projects such as Sustainability benchmarking for the South African Tourism Industry, Social ecology of conservation areas, and Climate change and tourism amongst others.

Raoul Merlin Ndonbou is a PhD holder since 2024 in Department of Earth Sciences of the University of Dschang. His work was focused on the Assessment of susceptibility factors and mapping of areas at risk of mass movement along the southern escarpment of the Bamileke plateau (west Cameroon). He obtained his MSc degree in 2018 in the same institution. He is specialised in Geotechnics and Environment, more specifically on natural hazards and related risks. He is the author of a scientific article published in an international journal and several communications presented in national and international conferences.

David Guimolaire Nkouathio is Professor at the Department of Earth Sciences at the University of Dschang since 2022. He holds a PhD from the University of Yaoundé 1 in 2006 under the theme Tectono-magmatic and volcanological evolution of the Cameroon Line: comparison of graben volcanism (Tombel Plain) and horst volcanism (Bambouto Mountains). His research area includes Volcanology, Petrology, and Geochemistry through which he has supervised and continues to supervise many master's and PhD students in the same institution. He is the author of several articles and member of scientific associations.

Marian Selorm Sapah is Senior Lecturer, with research interests focused in the areas of geochemistry, cosmochemistry, and geoscience education. She is an active member of several professional associations including the Ghana Institute of Geoscientists (GhIG) and the Geochemical Society, where she currently serves as the Public Affairs Officer and Board member, respectively. She enjoys mentoring, educational outreach, and public speaking and has a special interest in creating interest in and promoting geoscience as well as planetary and space science in Ghana and Africa. She is an advocate for the environment and sustainability as well as for women and girls in STEM.

Henry Bikwibili Tantoh is Senior Lecturer and a researcher at Department of Geography and Planning, the University of Bamenda. He holds a PhD in Geography and Environmental Studies from the University of the Witwatersrand South Africa, an MPhil and a BSc in Geography and Environmental Planning, University of Dschang-Cameroon. His research explores the socio-economic and political aspects of environmental management/governance and their links to wider societal change. Many of his works focus on politics of natural resource management in sub-Sahara Africa, with a particular focus on co-management of water resources and sustainability. He is also interested in climate change adaptation and climate-land- energy-food security nexus. Dr. Tantoh is Review Editor in Environmental Citizen Science and Guest Editor for Land, Livelihood and Food Security, and in Climate Services in Frontiers in Sustainable Food Systems.

Ghislain Zangmo Tefogoum is Associate Professor in the Department of Earth Sciences of the University of Maroua since May 2023. He obtained his bachelor's, master's and PhD degrees in Geology at the University of Dschang. He previously worked for the University of Dschang as a Teaching and Research Associate (2010–2012), for the University of Maroua as an Assistant (2012–2017) and Senior Lecturer (2017–2023). His PhD research was based on the volcanological characterisation of some calderas of the volcanic line of Cameroon. Study of the geoheritage and assessment of associated natural hazards: Case of the calderas of Manengouba, Bambouto and Bamenda Mountains. His main majors are geoheritage, and geomorphological azards and related risks. He has published various articles in peer-reviewed journals. In addition, he is the President of the Cameroon Association of Geomorphologists (CAG) since 2018 and a member of the Executive Committee of the International Association of Geomorphologists (IAG) since 2017. He was awarded the prestigious Brunsden Medal by the IAG in September 2022.

1 Geotourism, social sustainability and ecological justice in sub-Saharan Africa

Khodani Matshusa and Llewellyn Leonard

Introduction

Tourism in sub-Saharan Africa (SSA) has long been synonymous with wildlife safaris, drawing visitors from across the globe to witness the majestic creatures that roam its savannahs and forests. However, beneath the surface of this popular narrative lies a wealth of geological marvels waiting to be discovered and appreciated. Despite the continent's abundance of geological treasures, they have largely remained overshadowed by the allure of its wildlife. This disparity stems from a notable lack of research and attention devoted to geoheritage, geotourism, and the establishment of geoparks in the Global South, particularly in Africa, in comparison to the Global North. Scholars such as Matshusa and Leonard (2023) and Quesada-Valverde and Quesada-Román (2023) have highlighted this academic gap, underscoring the need for greater exploration and promotion of Africa's geological wonders. A glance at the scholarly landscape reveals a stark contrast in the distribution of geotourism research. The bulk of academic inquiry in this field is concentrated in regions like Europe, where countries such as Italy and Poland have seen extensive study, followed by Asia, with China and Iran taking the lead, and South America, notably Brazil, receiving considerable attention (Ruban, 2015). However, despite Africa's vast geological diversity, research efforts have been comparatively scant. In 2018, for instance, while Europe accounted for the lion's share of geotourism studies, Africa lagged behind with a mere 19 published articles on the subject (Ólafsdóttir and Tverijonaite, 2018). This underrepresentation underscores a critical need to shift the spotlight towards Africa's geological heritage and explore its potential as a vibrant hub for geotourism initiatives.

Geoheritage and geotourism represent cutting-edge paradigms within the realms of geoscience and Earth sciences, emerging as vital conduits to foster a deeper connection between society and the natural environment (Ólafsdóttir and Tverijonaite, 2018). These conceptual frameworks have evolved with the aim of bridging the gap between humanity's utilisation of natural resources and its responsibility to preserve and promote environmental sustainability and ecological equity. Geoheritage, as an integral component of natural heritage, holds intrinsic value in preserving the geological legacy of our planet, while geotourism emerges as a potent force in societal advancement, facilitating the engagement of local communities with

DOI: 10.4324/9781032706764-1

their heritage. The significance of geotourism extends beyond mere economic gain, serving as a catalyst for innovative solutions that facilitate the integration of local populations with their geoheritage resources and cultural legacies (Matshusa et al., 2021). By fostering a sense of ownership and pride among communities, geotourism not only safeguards geoheritage sites but also ensures their sustainable utilisation for the benefit of future generations. Moreover, in rural areas particularly, geotourism assumes a pivotal role in promoting social sustainability and ecological justice, empowering local stakeholders and fostering a more inclusive approach to environmental stewardship. Indeed, the potential of geotourism as a tool for fostering greater stakeholder participation cannot be overstated. By providing a platform for community engagement and dialogue, geotourism empowers local inhabitants to actively contribute to the resolution of issues affecting their territories (Matshusa et al., 2021). In essence, geotourism emerges not only as a means of economic development but also as a catalyst for social cohesion, environmental conservation, and sustainable development, thereby paving the way for a more harmonious relationship between society and its geological heritage.

Africa boasts a wealth of geological heritage, yet the development and promotion of geotourism by national tourism departments have been lacking. SSA stands as a treasure trove of unparalleled geological diversity with global significance. Across its vast expanse, the region boasts iconic landmarks that showcase its geological heritage to the world. Among these, the Cradle of Humankind World Heritage Site in South Africa stands as a testament to humanity's origins, offering a window into our evolutionary past. In Central Africa, the Cameroon Volcanic Line paints a dramatic geological canvas, narrating the story of volcanic activity that has shaped the landscape over millennia. The geographic tapestry of SSA is as varied as it is breathtaking. From the stark expanse of the low-lying Lake Assal in Djibouti, renowned for its status as the lowest point in Africa, to the majestic silhouette of Mount Kilimanjaro in Tanzania, the highest peak on the continent and the largest free-standing mountain in the world, the region's topography is a testament to the dynamic forces that have sculpted it. While it is impossible to encapsulate the entirety of SSA's geoheritage and geotourism activities within a single narrative, this book endeavours to spotlight a selection of sites that epitomise the region's geological significance. These sites represent not only important national treasures but also transboundary geological wonders of international acclaim. Through exploration and appreciation of these sites, we gain not only a deeper understanding of our planet's geological history but also a renewed appreciation for the remarkable diversity that defines SSA.

Despite the continent's geological richness, local communities residing near these geoheritage sites continue to grapple with high unemployment rates and pervasive poverty. This book endeavours to offer an African-centric perspective on geotourism, departing from the predominantly Western viewpoints that have dominated the discourse thus far. As highlighted, the focus on Africa is imperative due to the paucity of research aimed at identifying and cataloguing geoheritage sites across the continent. As far as we know, this book represents a pioneering effort to underscore the role of geotourism as a catalyst for social sustainability and ecological justice in SSA. Its thematic breadth encompasses a range of critical issues, from leveraging geotourism for local socio-economic development to the intersection of geotourism with environmental conservation efforts. Moreover, it delves into the intricacies of geoheritage and geotourism threats in the face of climate change and extractive industries such as mining. Given its comprehensive scope, the book presents a unique African perspective that holds significant value for a diverse array of stakeholders. These include local communities seeking avenues for economic empowerment, community leaders striving for sustainable development, the private sector exploring investment opportunities, government departments tasked with tourism promotion, academics delving into geoscience and environmental studies, and individuals passionate about preserving geoheritage and fostering responsible tourism practices. By offering insights, strategies, and case studies tailored to the African context, this book aims to inspire informed action and collaboration among stakeholders, ultimately fostering a more inclusive and sustainable approach to geotourism development while ensuring social equity and environmental stewardship.

Concepts of geoheritage and geotourism: brief overview

The recognition of geoheritage sites and geotourism as catalysts for societal development and environmental conservation has reverberated across the globe in recent years. Geoheritage, as elucidated by Newsome and Dowling (2018), encompasses the elements of the Earth that we hold dear, constituting a vital component of our natural heritage. It encompasses a diverse array of geological phenomena, including mineralogical formations, geomorphological landscapes, traces of ancient life in the form of fossils, expansive cave systems, and the historical legacy of mining activities. These geoheritage sites are not merely geographical entities; they are repositories of invaluable natural treasures that bear witness to the Earth's dynamic evolution over millennia. Lima et al. (2010) emphasise the paramount importance of geoheritage sites on a global scale, underscoring their exceptional natural characteristics that warrant preservation and appreciation. Furthermore, geotourism emerges as a multifaceted concept that transcends mere tourist activities, as delineated by the International Congress of Geotourism (2011). It seeks to sustain and enrich the identity of a territory by leveraging its geological, environmental, cultural, aesthetic, and historical assets while prioritising the well-being of local residents. In essence, geotourism represents a holistic approach to tourism that integrates various facets of a destination's heritage, fostering sustainable development and fostering a deeper connection between visitors and the land they explore. Building upon these foundational definitions, Matshusa et al. (2021) offer a practical and locally grounded perspective on geotourism. They define it as the continual utilisation of geoheritage sites for tourism activities aimed at benefiting local communities. This pragmatic approach underscores the transformative potential of geotourism as a driver of economic empowerment and community development, positioning geoheritage sites not only as attractions for visitors but also as engines for positive change within their surrounding communities. The intertwined concepts of geoheritage and geotourism represent more than just avenues for leisure and recreation; they embody a profound commitment to preserving our planet's natural legacy while harnessing its potential to foster sustainable development and cultivate meaningful connections between people and place.

The concepts of geoheritage and geotourism are intimately linked to the burgeoning field of UNESCO global geoparks, representing a novel niche area within the realm of heritage conservation and sustainable development. As articulated by UNESCO, geoparks serve as cohesive geographical entities where sites and landscapes of international geological significance are managed through a holistic approach encompassing protection, education, and sustainable development (UNESCO, 2018). Despite their relatively recent emergence, the UNESCO Global Geoparks Network has swiftly expanded, comprising 195 geoparks across 48 countries as of January 2024. However, the representation of Africa within this network remains sparse, with only two geoparks – M'Goun in Morocco and Ngorongoro in Tanzania – bearing the UNESCO designation. In crafting the approach for this book, due consideration was given to both regional and international initiatives pertaining to geoheritage and geotourism. At the global level, UNESCO conventions such as the 1972 Convention concerning the Protection of the World Cultural and Natural Heritage and the 2003 Convention for the Safeguarding of the Intangible Cultural Heritage serve as guiding frameworks for the legal recognition and preservation of sites of international significance (UNESCO, 1972a, 1972b). These conventions lay the groundwork for the identification, conservation, and promotion of exceptional geoheritage sites and geoparks worldwide. Moreover, UNESCO emphasises the rights of indigenous communities to access, manage, and derive benefits from their natural heritage, reinforcing principles of inclusivity and cultural preservation (UNESCO, 2003). Despite the rich historical and cultural tapestry of SSA, the region's geological sites of international significance remain relatively understudied and underrepresented within the global geotourism discourse. From an African perspective, there is a notable dearth of coordinated research and programmes focusing on geoheritage and geotourism, exemplified by the limited presence of UNESCO geoparks on the continent. The African Geoparks Network, in particular, has yet to make significant inroads into the global geotourism narrative (Matshusa and Leonard, 2023). Against this backdrop, the individual chapters of this book serve as illuminating case studies, spotlighting key geoheritage and geotourism research initiatives within SSA countries. By showcasing the diverse geological wonders and cultural heritage of the region, this book aims to catalyse greater attention and appreciation for Africa's unique contribution to the global geological landscape. Through collaborative efforts and informed discourse, it is hoped that the potential of geoheritage and geotourism in SSA can be fully realised, fostering sustainable development and cultural enrichment for generations to come.

Geotourism has emerged as a powerful mechanism for fostering sustainable tourism development while simultaneously promoting geoconservation and minimising extractive and harmful activities. Beyond simply serving as a means for tourists to bask in the natural beauty of an area, geotourism offers a multifaceted experience that enriches both visitors and local communities alike. As noted by Crofts et al. (2021), geotourism provides a unique opportunity for individuals to delve into the intricate world of geological processes, gaining insight into the formation of landscapes and the importance of preserving their geological heritage. However, the successful implementation of geotourism hinges upon careful planning and assessment. Prior to activating geotourism initiatives, it is imperative to conduct comprehensive inventories and sensitivity analyses of geoheritage sites (Lima et al., 2010). These studies serve as the foundation for developing geotourism routes and determining the carrying capacity of destinations, thereby ensuring that tourism activities remain sustainable and minimise negative impacts on the environment. Moreover, geotourism presents a unique opportunity for the Global South, particularly SSA, to integrate geoheritage sites with historical and cultural landmarks, both in rural areas and within urban centres (Matshusa et al., 2021). By leveraging the intrinsic linkages between geological, historical, and cultural heritage, geotourism initiatives can serve as catalysts for economic development, community empowerment, and cultural preservation. In doing so, they not only showcase the natural wonders of the region but also highlight its rich tapestry of history and culture, fostering a deeper appreciation for the interconnectedness of humanity and the environment. Thus, geotourism represents a paradigm shift in tourism management, offering a holistic approach that prioritises sustainability, education, and cultural enrichment. By harnessing the potential of geoheritage sites and integrating them with broader historical and cultural narratives, geotourism has the power to transform destinations, empower local communities, and foster a greater sense of stewardship towards our planet's geological heritage. One of the most significant advantages of geotourism lies in its potential to mitigate human-wildlife conflict and promote coexistence by championing environmentally friendly activities, particularly within protected areas. These areas play a pivotal role not only in biodiversity conservation but also in driving economic development through sustainable tourism practices such as geotourism (Gupta et al., 2023).

By fostering a harmonious relationship between local communities, broader society, and wildlife, geotourism offers a unique platform for promoting both conservation efforts and socio-economic development. Moreover, geotourism holds promise as a proactive response to threats facing geoheritage sites, including the impacts of climate change and mining activities. While it is widely acknowledged for its potential to benefit conservation efforts, foster appreciation through interpretation of geological heritage, and stimulate economic growth (Newsome and Dowling, 2018), there exists a crucial missing link – the social component. This component necessitates a deeper consideration of the benefits and meaningful participation of local communities in the geotourism value chain. Therefore, we propose that geotourism should be viewed as a transdisciplinary paradigm, cutting across not only geosciences but also social sciences, natural sciences, and environmental studies. As depicted in Figure 1.1, geotourism yields a myriad of benefits, including enhancing the well-being of local communities, stimulating local economic development, diversifying market opportunities, fostering awareness and knowledge generation, and promoting geoconservation. By embracing

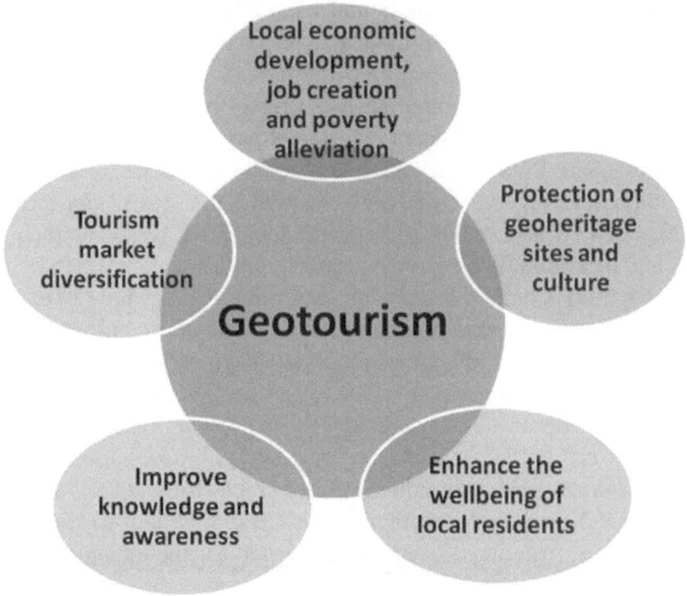

Figure 1.1 The benefits of geotourism (Matshusa, 2020).

a transdisciplinary approach, this introductory chapter underscores the multifaceted nature of geotourism, emphasising its role in promoting social sustainability and ecological justice frameworks. Geotourism transcends traditional disciplinary boundaries, offering a holistic framework that integrates environmental, social, and economic considerations. By recognising and harnessing the interconnectedness of these domains, geotourism emerges as a powerful tool for promoting sustainable development, preserving natural heritage, and fostering equitable engagement among diverse stakeholders.

Studies have underscored the dearth of pro-poor economic growth and development within the tourism sector in SSA (Akama and Kieti, 2007; Saayman et al., 2012; Chirenje et al., 2013). However, research suggests that nature-based tourism, including geotourism, holds the potential to yield significant economic benefits for local communities (Sagala et al., 2018; Gupta et al., 2023). While geotourism remains relatively underdeveloped in SSA, its geoheritage sites harbour immense potential to drive local economic development. These sites offer various avenues for economic empowerment, including income generation, employment opportunities, ownership stakes, tour operation and guiding services, provision of goods and services, accommodation, and retail businesses. Numerous studies conducted across different continents have highlighted the diverse ways in which geotourism can foster sustainable local economic development, ranging from Asia (Lee and Jayakumar, 2021) to Europe (Farsani et al., 2011), Oceania (Dowling, 2013), and the USA (Stokes et al., 2003). However, from the perspective of SSA, limited research exists on the economic impacts of geotourism, with studies scattered across various countries including Angola (Henriques et al., 2013), Egypt (Sallam et al., 2018), Ethiopia (Tesfa and Zewdie, 2023), Morocco (Enniouar et al., 2015), Namibia (Dowling and Pforr, 2021), South Africa (Matshusa et al., 2021), and Zambia (Ngwira, 2015). This scarcity of research underscores the missed opportunities for SSA countries to fully leverage the potential of geotourism for local economic development. To address this gap, SSA nations must devise strategies aimed at harnessing the economic potential of their exceptional geological resources. This can entail identifying and formally recognising geoheritage sites with international significance, potentially designating them as UNESCO geoparks. Indeed, geoparks have demonstrated their capacity to catalyse local economic development (Ruban et al., 2023). This book delves into the interventions, activities, and potential economic benefits that could emanate from SSA's remarkable geological wealth, providing valuable insights and strategies for unlocking the economic potential of geotourism in the region. The concept of sustainable development generally seeks to emphasise the balance among the environmental, economic, and social initiatives. The meaning and associated research of the social pillar remain limited. Like geotourism, social sustainability has suffered marginalisation in the global sustainable development agenda and literature (Weingaertner and Moberg, 2014; Barron et al., 2023).

Previous studies have shown the vagueness and lack of research of the social sustainability pillar, thereby hindering the objective of achieving the balance with the economic and environmental sustainability pillars (see Agyeman et al., 2002; Cuthill, 2010; Dempsey et al., 2011; Murphy, 2012; Vifell and Soneryd, 2012; Partridge, 2014; Taherkhani, 2023). As such, a traditional sustainability agenda tends to result in the exclusion of social issues such as inequality and participation in decision-making processes by local communities (Agyeman et al., 2002). If not addressed, these deficiencies can result in policy failure. To address this gap, Taherkhani (2023) recommends five subsystems of social sustainability, thus, socio-political, socio-cultural, socio-institutional, socio-economic and socio-environmental. Cuthill (2010) suggests that key components of the social sustainability framework include social capital, social infrastructure, social justice and equity and engaged governance. Furthermore, a social sustainability framework should include development sustainability, maintenance sustainability, and bridge sustainability (Vallance et al., 2011). Development sustainability seeks to address basic needs, equity, employment, education, justice, freedom, access to influential decision-making and general capacity-building, the distribution of power and resources, and access to basic infrastructure and services.

Maintenance sustainability includes the maintenance of socio-cultural features and how people embrace or resist change. As such, it is proper to argue that economic and environmental aspects of sustainability are highly reflected in a social sustainability framework (Cuthill, 2010). Murphy (2012) suggests that the social pillar of sustainability should be expanded to include equity, awareness of sustainability, participation, and social cohesion. In addition, geotourism has the potential to ensure involvement of the local community in decision-making processes and in the promotion and appreciation of local culture because of the local community's special interests in each geotourism activity (Ngwira, 2015; Tavares et al., 2015). Given that geotourism considers the well-being of local communities abutting geo-heritage sites, it is critical to understand the social sustainability framework. In this regard, Becker and Jahn (1999) defined social sustainability as long-term relationships among nature and society that lead to the viability of society. Therefore, impacts of tourism on natural resources, cultural resources and host communities must also be monitored using social sustainability indicators. These indicators will test the effectiveness of the geotourism programme to contribute to social sustainability and environmental justice. Barron et al. (2023) argue that social sustainability should be integrated into the development discourse in theory and practice. In this case, Matshusa (2020) suggested seven social sustainability indicators in relation to geotourism (Figure 1.2).

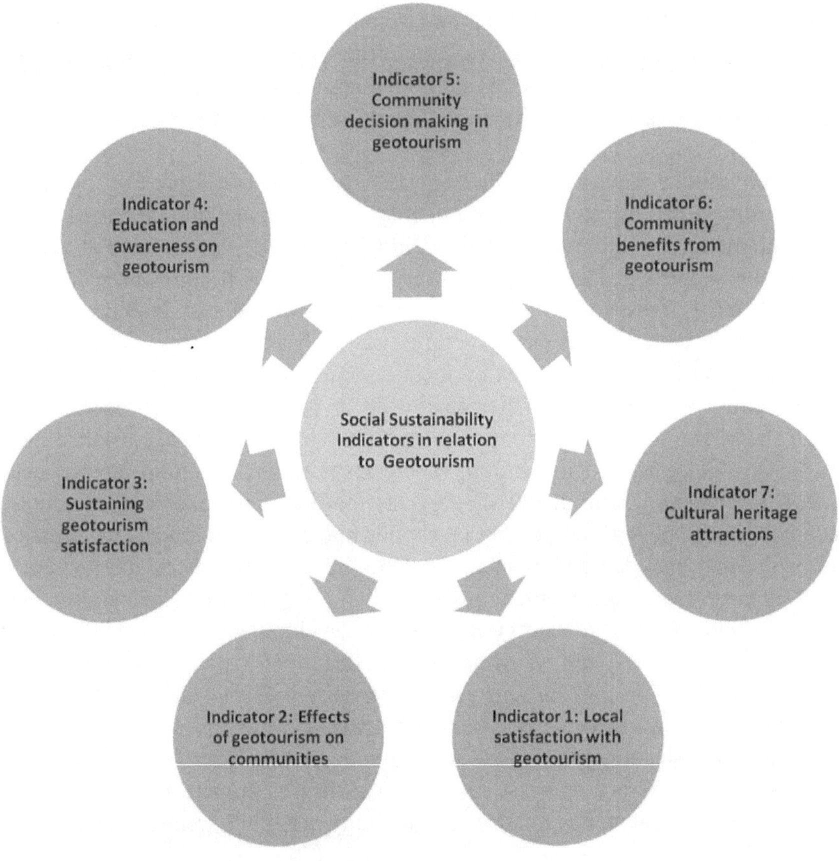

Figure 1.2 Framework for social sustainability indicators in relation to geotourism (Matshusa, 2020).

Some authors and scholars believe that the concept of social sustainability should be identified autonomously, while others argue that it should be defined in relation to other pillars of sustainability, mainly the notion of environmental sustainability (Rasouli and Kumarasuriyar, 2016). In this context, social sustainability policies embrace environmental sustainability (Ballet et al., 2020). Nature and the environment have continued to receive international attention as part of the sustainable development agenda. Environmental sustainability refers to the utilisation of the natural resources and maintenance of the environmental assets to provide income without depleting the natural capital (Goodland, 1995). Natural capital represents the environmental and natural resource assets of development and preservation (Jabareen, 2008). In this context, geotourism caters for the utilisation of natural assets such as geoheritage sites. Furthermore, social sustainability is directly linked with ecological justice. Ecological justice – or environmental justice – has gained prominence in recent years owing to negative anthropogenic activities on the environment and society. Ecological justice was designed to ensure that local communities' voices are heard, especially concerning the protection and beneficiation of their heritage resources. Environmental justice is defined as "the fair treatment and meaningful involvement of all people regardless of race, color, national origin, or income with respect to the development, implementation, and enforcement of laws, regulations and policies that affect the environment and/or public health" (First National People of Color Environmental Leadership Summit, 1991; Lee, 2021). Therefore, there is a need for community members and other key stakeholders to participate in decision-making processes, including research concerning their environment and development of sustainable interventions and implementation strategies to address them (Van Horne et al., 2023).

The integration of geotourism with social sustainability and ecological justice can provide a unique opportunity for solving some of the major societal challenges in the continent such as high unemployment levels, poverty, lack of community participation, climate change impacts, and other environmental injustices. Despite its transdisciplinary nature, there remains a noticeable dearth of literature addressing the threats posed to geoheritage and geotourism by both anthropogenic and natural activities. Among these threats, mining and climate change loom large as major global concerns, capable of undermining the integrity and sustainability of destinations, particularly jeopardising the scientific, social, and economic benefits associated with geoheritage sites and geotourism. Recognising the urgency of this issue, this book endeavours to identify and confront some of the most pressing threats facing geoheritage sites and geotourism through a series of insightful case studies within SSA. By delving into real-world scenarios, the chapters illuminate the multifaceted challenges posed by mining activities, environmental degradation, and other anthropogenic impacts, as well as the escalating risks posed by climate change. Moreover, in addition to diagnosing these threats, the book puts forth a range of strategic recommendations aimed at enhancing geoconservation efforts and safeguarding the invaluable heritage of geoheritage sites. By offering practical solutions grounded in both scientific rigour and community engagement, the chapters provide a roadmap for policymakers, conservationists, and stakeholders to mitigate the risks facing geoheritage and geotourism in SSA. In essence, this book serves as a vital resource for understanding and addressing the complex interplay between human activities, environmental dynamics, and the preservation of geological heritage. Through informed analysis and proactive intervention, it seeks to ensure the long-term sustainability and resilience of geoheritage sites and geotourism initiatives in the face of mounting threats.

Book structure overview

This book is a symphony of knowledge and insight, meticulously orchestrated to unravel the intricate connections between social sustainability, ecological justice, and the enchanting realm of geoheritage sites and geotourism within SSA. Authored by a diverse array of scholars hailing from the SSA region, its chapters serve as windows into a world where the preservation of natural and cultural heritage intertwines with the pursuit of equitable and sustainable development. At its heart, the book serves as a beacon of enlightenment, illuminating the transformative potential of geotourism as a vehicle for fostering social cohesion and environmental stewardship across SSA. Through a lens finely calibrated to capture the nuances of the region's socio-ecological landscape, it navigates the complex terrain of challenges and opportunities, offering a roadmap towards a more resilient and equitable future. Each chapter is a testament to the power of interdisciplinary inquiry, weaving together strands of empirical research, theoretical frameworks, and case studies to paint a holistic portrait of the role of geotourism in promoting sustainable development within SSA. This book has been arranged in such a way as to link social sustainability and ecological justice with geoheritage sites and geotourism within the SSA. It consists of several chapters written by authors from the SSA region.

In Chapter 2, Donkor, Tantoh, and Mearns explore the role of geotourism for local economic development in South Africa and look at the constraints and opportunities. Chapter 3, by Dyssel, conducts a social-ecological analysis of geoheritage resource uses and the livelihood concerns of villagers in the Cederberg Mountains, Western Cape, South Africa. Chapter 4, by Matshusa, examines the impacts of climate change on geoheritage and geotourism in Southern Africa. In Chapter 5, Leonard conducts a critical examination of the political economy of mining development impacts for sustainable development and geotourism in particular. In Chapter 6, Tantoh, Donkor, and McKay examine the potential contribution of geotourism and environmental protection for sustainable development in Cameroon. In Chapter 7, Tefogoum, Dedzo, Dama, Ndonbou, Dongmo, and Nkouathio explore a geological heritage for geotouristic and geoeducational practices at the Mount Bambouto slopes geomorphosites in Cameroon. The chapter examines the region as a diversifying tourism activity and beyond the country's substantial natural resources such as the ecological, cultural, religious and historic types. In Chapter 8, Sapah, Asiedu, Loh, and Banoeng-Yakubo examine geoconservation strategies to enhance geoscience training in tertiary institutions in Ghana. The book's case studies offer glimpses into the diverse array of challenges and opportunities that characterise the region's geotourism landscape. Through analysis and thought-provoking insights, the book provides the possibilities for sustainable development within SSA and stands as a beacon of inspiration to the transformative potential of geotourism as a force for positive change. It offers a glimpse into a future where social sustainability, ecological justice, and geoheritage conservation converge to create a more equitable and sustainable world for generations to come.

References

Agyeman, J., Bullard, R.D. and Evans, B. (2002). Exploring the Nexus: Bringing Together Sustainability, Environmental Justice and Equity. *Journal of Space and Polity*, 6(1), pp. 77–90. http://doi.org/10.1080/13562570220137907.

Akama, J.S. and Kieti, D. (2007). Tourism and Socio-Economic Development in Developing Countries: A Case Study of Mombasa Resort in Kenya. *Journal of Sustainable Tourism*, 15(6), pp. 735–748. https://doi.org/10.2167/jost543.0.

Ballet, J., Bazin, D. and Mahieu, F.G. (2020). A Policy Framework for Social Sustainability: Social Cohesion, Equity and Safety. *Sustainable Development*, 28(5), pp. 1388–1394. https://doi.org/10.1002/sd.2092.

Barron, P., Cord, L., Cuesta, J., Espinoza, S.A., Larson, G. and Woolcock, M. (2023). *Social Sustainability in Development: Meeting the Challenges of the 21st Century. New Frontiers of Social Policy*. Washington, DC: World Bank.

Becker, E. and Jahn, T. (1999). *Sustainability and the Social Sciences*. New York, USA: Zed Books.

Chirenje, L.I., Chitotombe, J., Gukurume, S., Chazovachii, B. and Chitongo, L. (2013). The Impact of Tourism Leakages on Local Economies: A Case Study of Nyanga District, Zimbabwe. *Journal of Human Ecology*, 42, pp. 9–16. https://doi.org/10.1080/09709274.2013.11906576.

Crofts, R., Tormey, D. and Gordon, J.E. (2021). Introducing New Guidelines on Geoheritage Conservation in Protected and Conserved Areas. *Geoheritage*, 13, pp. 1–14. https://doi.org/10.1007/s12371-021-00552-0.

Cuthill, M. (2010). Strengthening the Social in Sustainable Development: Developing a Conceptual Framework for Social Sustainability in a Rapid Urban Growth Region in Australia. *Sustainable Development*, 18(6), pp. 362–373. https://doi.org/10.1002/sd.397.

Dempsey, N., Bramley, G., Power, S. and Brown, C. (2011). The Social Dimension of Sustainable Development: Defining Urban Social Sustainability. *Sustainable Development*, 19(5), pp. 289–300. https://doi.org/10.1002/sd.417.

Dowling, R.K. (2013). Global Geotourism – An Emerging Form of Sustainable Tourism. *Czech Journal of Tourism*, 2(2), pp. 59–79. https://doi.org/10.2478/cjot-2013-0004.

Dowling, R.K. and Pforr, C. (2021). Geotourism – A Sustainable Development Option for Namibia. *Journal of Ecotourism*, 20(4), pp. 371–385. https://doi.org/10.1080/14724049.2021.1910699.

Enniouar, A., Errami, E., Lagnaoui, A. and Bouaala, O. (2015). The Geoheritage of the Doukkala-Abda Region (Morocco): An Opportunity for Local Socio-Economic Sustainable Development. In: E. Errami, M. Brocx and V. Semeniuk (Eds.), *From Geoheritage to Geoparks, Case Studies from Africa and Beyond* (pp. 109–124). https://doi.org/10.1007/978-3-319-10708-0_7.

Farsani, N.T., Coelho, C.O.A. and Costa, C.M.M. (2011). Geotourism and Geoparks as Novel Strategies for Socio-economic Development in Rural Areas. *International Journal of Tourism Research*, 13(1), pp. 68–81. https://doi.org/10.1002/jtr.800.

First National People of Color Environmental Leadership Summit. (1991, October). The Principles of Environmental Justice (EJ). https://www.ejnet.org/ej/principles.pdf (Accessed 25 May 2021).

Goodland, R. (1995). The Concept of Environmental Sustainability. *Annual Review of Ecology and Systematics*, 26, pp. 1–24. https://doi.org/10.1146/annurev.es.26.110195.000245.

Gupta, A., Zhu, H., Bhammar, H., Earley, E., Filipski, M., Narain, U., Spencer, P., Whitney, E. and Taylor, J.E. (2023). Economic Impact of Nature-Based Tourism. *PLoS ONE*, 18(4), pp. 1–16. https://doi.org/10.1371/journal.pone.0282912.

Henriques, M.H., Tavares, A.O. and Bala, A.L.M. (2013). The Geological Heritage of Tundavala (Angola): An Integrated Approach to its Characterisation. *Journal of African Earth Sciences*, 88, pp. 62–71. http://doi.org/10.1016/j.jafrearsci.2013.09.003.

International Congress of Geotourism. (2011). *Arouca Declaration on Geotourism*.www.europeangeoparks.org/?p=223 (Accessed 3 February2022).

Jabareen, Y. (2008). A New Conceptual Framework for Sustainable Development. *Environment Development Sustainability*, 10(2), pp. 179–192. https://doi.org/10.1007/s10668-006-9058-z.

Lee, C. (2021). Evaluating Environmental Protection Agency's Definition of Environmental Justice. *Environmental Justice*, 14(5), pp. 332–337. https://doi.org/10.1089/env.2021.0007.

Lee, Y. and Jayakumar, R. (2021). Economic Impact of UNESCO Global Geoparks on Local Communities: Comparative Analysis of Three UNESCO Global Geoparks in Asia. *International Journal of Geoheritage and Parks*, 9, pp. 189–198. https://doi.org/10.1016/j.ijgeop.2021.02.002.

Lima, F.F., Brilha, J.B. and Salamuni, E. (2010). Inventorying Geological Heritage in Large Territories: A Methodological Proposal Applied to Brazil. *Geoheritage*, 2(3–4), pp. 91–99. https://doi.org/10.1007/s12371-010-0014-9.

Matshusa, K. (2020). The Potential for Geotourism at the Kruger National Park for Social Sustainability. PhD Thesis, University of Johannesburg.

Matshusa, K. and Leonard, L. (2023). Proposed Governance Model for Geotourism and Geoheritage in South Africa and the Implications for Africa. *Geoheritage*, 15, pp. 1–18. https://doi.org/10.1007/s12371-022-00783-9.

Matshusa, K., Leonard, L. and Thomas, P. (2021). The Contribution of Geotourism to Social Sustainability: Missed Opportunity? *The International Journal of Social Sustainability in Economic, Social, and Cultural Context*, 17, pp. 95–118. https://doi.org/10.18848/2325-1115/CGP/v17i01/95-118.

Murphy, K. (2012). The Social Pillar of Sustainable Development: A Literature Review and Framework for Policy Analysis. *Sustainability: Science, Practice and Policy*, 8(1), pp. 15–29. https://doi.org/10.1080/15487733.2012.11908081.

Newsome, D. and Dowling, R. (2018). Geoheritage and Geotourism. In: E. Reynard and J. Brilha (Eds.), *Geoheritage – Assessment, Protection and Management*. Amsterdam: Elsevier. https://doi.org/10.1016/B978-0-12-809531-7.00017-4.

Ngwira, P.M. (2015). Geotourism and Geoparks: Africa's Current Prospects for Sustainable Rural Development and Poverty Alleviation. In: E. Errami, M. Brocx and V. Semeniuk (Eds.), *From Geoheritage to Geoparks, Case Studies from Africa and Beyond* (pp. 25–33). https://doi.org/10.1007/978-3-319-10708-0_2.

Ólafsdóttir, R. and Tverijonaite, E. (2018). Geotourism: A Systematic Literature Review. *Geosciences*, 8(7), pp. 1–16. https://doi.org/10.3390/geosciences8070234.

Partridge, E. (2014). Social Sustainability. In: A.C. Michalos (Ed.), *Encyclopedia of Quality of Life and Well-Being Research* (pp. 6178–6186). https://doi.org/10.1007/978-94-007-0753-5_2790.

Quesada-Valverde, M.E. and Quesada-Román, A. (2023). Worldwide Trends in Methods and Resources Promoting Geoconservation, Geotourism, and Geoheritage. *Geosciences*, 13(2), pp. 1–12. https://doi.org/10.3390/geosciences13020039.

Rasouli, A.H. and Kumarasuriyar, A. (2016). The Social Dimension of Sustainability: Towards Some Definitions and Analysis. *Journal of Social Science for Policy Implications*, 4(2), pp. 23–34. https://doi.org/10.15640/jsspi.v4n2a3.

Ruban, D.A. (2015). Geotourism – A Geographical Review of the Literature. *Tourism Management Perspectives*, 15, pp. 1–15. https://doi.org/10.1016/j.tmp.2015.03.005.

Ruban, D.A., Mikhailenko, A.V., Yashalova, N.N. and Scherbina, A.V. (2023). Global Geoparks: Opportunity for Developing or "Toy" for Developed? *International Journal of Geoheritage and Parks*, 11, pp. 54–63. https://doi.org/10.1016/j.ijgeop.2022.11.003.

Saayman, M., Rossouw, R. and Krugell, W. (2012). The Impact of Tourism on Poverty in South Africa. *Development Southern Africa*, 29(3), pp. 462–487. https://doi.org/10.1080/0376835X.2012.706041.

Sagala, S., Rosyidie, A., Sasongko, M.A. and Syahbid, M.M. (2018). Who Gets the Benefits of Geopark Establishment? A study of Batur Geopark Area, Bali Province, Indonesia. *IOP Conference Series: Earth and Environmental Science*, 158, pp. 1–16. https://doi.org/10.1088/1755-1315/158/1/012034.

Sallam, E.S., Ponedelnik, A.A., Ties, G., Yashalova, N.N. and Ruban, D.A. (2018). The Geological Heritage of the Kurkure Dungul Area in Southern Egypt. *Journal of African Earth Sciences*, 137, pp. 103–115. https://doi.org/10.1016/j.jafrearsci.2017.10.012.

Stokes, A.M., Cook, S. and Drew, D. (2003). *Geotourism: The New Trend in Travel*. www.egret.us/clinton/Geotourism%20The%20New%20Trend%20in%20Travel.pdf (Accessed 18 September2017).

Taherkhani, R. (2023). An Integrated Social Sustainability Assessment Framework: The Case of Construction Industry. *Open House International*, 48(2), pp. 206–236. https://doi.org/10.1108/OHI-04-2022-0098.

Tavares, A.O., Henriques, M.H., Domingos, A. and Bala, A. (2015). Community Involvement in Geoconservation: A Conceptual Approach Based on the Geoheritage of South Angola. *Sustainability*, 7(5), pp. 4893–4918. https://doi.org/10.3390/su7054893.

Tesfa, C. and Zewdie, M.M. (2023). Assessment and Map of Geotourism Potential Sites in Amhara Regional State, Northwestern Ethiopia. *International Journal of Geoheritage and Parks*, 11, pp. 634–651. https://doi.org/10.1016/j.ijgeop.2023.11.005.

UNESCO. (1972a, 17 October–21 November). *Resolutions, Recommendations, Volume 1: Records of the General Conference, Seventeenth Session*. Paris. http://unesdoc.unesco.org/images/0011/001140/114044E.pdf (Accessed 8 February 2017).

UNESCO. (1972b, 16 November). *Convention Concerning the Protection of the World Cultural and Natural Heritage: Adopted by the General Conference at Its Seventeenth Session*. Paris. http://whc.unesco.org/archive/convention-en.pdf (Accessed 8 February2017).

UNESCO. (2003). *2003 Convention for the Safeguarding of the Intangible Cultural Heritage*. https://ich.unesco.org/doc/src/18440-EN.pdf (Accessed 22 January 2024).

UNESCO. (2018). *Global Geoparks*. www.unesco.org/new/en/natural-sciences/environment/earth-sciences/unesco-global-geoparks/ (Accessed 11 June2018).

Vallance, S., Perkins, H.C. and Dixon, J.E. (2011). What is Social Sustainability? A Clarification of Concepts. *Geoforum*, 42(3), pp. 342–348. https://doi.org/10.1016/j.geoforum.2011.01.002.

Van Horne, Y.O., Alcala, C.S., Peltier, R.E., Quintana, P.J.E., Seto, E., Gonzales, M., Johnston, J.E., Montoya, L.D., Quirós-Alcalá, L. and Beamer, P.I. (2023). An Applied Environmental Justice Framework for Exposure Science. *Journal of Exposure Science and Environmental Epidemiology*, 33, pp. 1–11. https://doi.org/10.1038/s41370-022-00422-z.

Vifell, A.C. and Soneryd, L. (2012). Organizing Matters: How the Social Dimension Gets Lost in Sustainability Projects. *Sustainable Development*, 20(1), pp. 18–27. https://doi.org/10.1002/sd.461.

Weingaertner, C. and Moberg, A. (2014). Exploring Social Sustainability: Learning From Perspectives on Urban Development and Companies and Products. *Sustainable Development*, 22(2), pp. 122–133. https://doi.org/10.1002/sd.536.

2 Exploring the potential role of geotourism in local economic development in South Africa

Opportunities and constraints

Felix Ekwabena Donkor, Henry Bikwibili Tantoh, and Kevin Mearns

Introduction

A quest for global sustainable development

The quest for progress has been a key feature of human communities since time immemorial. This is reflected in historical accounts of civilisations and humanity's harnessing of environmental resources to support its development. Contemporary society, however, has begun focusing to an increasing degree on the concept of sustainable development. A set of Sustainable Development Goals (SDGs), also referred to as the Global Goals, has therefore been accepted by the United Nations (UN) member states as a global rallying point for activities directed towards eradicating poverty while safeguarding the planet UNDP, 2023). It is noteworthy that the 17 SDGs are mutually interactive, in that interventions in one area have an effect in others. Moreover, they emphasise that meaningful development needs to harmonise social, economic and environmental imperatives for sustainability. One of the hallmarks of the SDGs is the commitment by UN member states to foregrounding the progress of individuals and communities regarded as being furthest behind (UNDP, 2023). In order for the global sustainable development agenda to be realised, it is vital to harness the creativity, innovation and enterprise of the entire global citizenry for the success of all facets of the SDGs.

Regional bodies such as the African Union (AU) have developed parallel schemes to promote sustainable development, as encapsulated in the African Union's Agenda 2063, or Vision 2063. Although Africa has repeatedly faced obstacles such as conflict, instability, and poverty that have undermined sustainable development on the continent, African countries continue to strive for improvement (African Union, 2023). Thus, Vision 2063, or Agenda 2063: The Africa We Want, is seen as a blueprint for Africa's transformation into an influential player in the global community, embodying the continent's strategic plans for inclusive and sustainable development, progress and collective prosperity (African Union, 2023). South Africa has been one of the prime architects of the AU's Agenda 2063 framework, and so it is not surprising that some aspects have filtered through into South Africa's National Development Plan (NDP), or Vision 2030.

DOI: 10.4324/9781032706764-2

South Africa's NDP is a national action plan that seeks to provide a decent standard of living for all South Africans by 2030, among other things through the eradication of poverty and the reduction of inequality. It is envisaged that these objectives can be achieved through harnessing human capital, nurturing a broad inclusive economy, capacity development, building the capabilities of the state, and advancing good governance and collaboration throughout society. The NDP thus represents a blueprint for enhancing South Africa's future. This involves dealing with the factors causing poverty and inequality by shifting the pivot of policy formulation from short-term to longer-term measures based on solid evidence. A common thread in the UN SDGs, the AU Agenda 2063, and South Africa's Vision 2030 is that governments on their own are unable to meet the multi-faceted socio-economic needs of society. In order for the gains of development to be felt at the local level, the involvement of all social actors and partners across the social divide is necessary for meaningful and inclusive sustainable development (Fourie, 2018).

Exploring avenues for inclusive development at the grassroots: the notion of local development

Mensah et al. (2017) suggest that local development has gained global currency as a robust alternative to top-down development measures, which have often been unable to create meaningful and sustainable development at the local level. This is particularly so in light of the argument that globalisation has altered the factors that control the world's economies, joining national, regional, and local economies in an unprecedented and dynamic way (Rogerson & Rogerson, 2010). Furthermore, local economies are increasingly subject to measures introduced at the supranational level. Local development is more regional, with the resources and endogenous elements of an area being central to its socio-economic development. In addition, it entails the development of an area being led by people and entities who are based there and whose undertakings there have significant impact (Kisman & Tasar, 2014). This involves policy formulation, consensus building, and the rolling out of interventions that permit civic engagement in working to overcome a myriad of socio-economic challenges so as to promote sustainable development and enhance a society's quality of life or standard of living (Agbevade, 2020).

It has been observed that there is a close connection between investment in resources and the development of an area (Doan, 2022). However, it has also been shown that the gains of economic development do not always reach those at the bottom of the pyramid or the margins of society. This makes the consideration of viable measures for locally relevant socio-economic development imperative (Agbevade, 2020), since investment in the harnessing of local resources leads to employment creation and the enhancement of livelihoods (Doan, 2022). A number of scholars are nevertheless of the view that the notion of local development is either elusive or contested, since there are so many definitions of the concept (Mensah et al., 2017). For example, some scholars refer to local development as an expansion of the capabilities of a local economy to create wealth for the community, while others define it as a way of stimulating the local economy to expand and generate employment opportunities, in particular by harnessing community-based resources (Diko et al., 2021). A common denominator in the narrative as it relates to local development, however, is the need to forge partnerships among core participants, such as local officials, non-governmental organisations, community organisations, and civic groups, for the purpose of creating the requisite conducive environment for economic growth and employment generation (Agbenyefia, 2022; Adamowicz, 2023). The concept of local development features prominently in South Africa's NDP. In Vision 2030 it is contended that many rural communities currently face economic hardship, and are really nothing more than migrant sending centres or sources of human capital for the urban areas. Vision 2030 therefore seeks to transform South Africa's rural areas into successful economies buoyed by vibrant local development that will significantly enhance the living standards of the rural populace. A number of approaches are being explored to facilitate the local development espoused in the NDP. One such approach is geotourism.

Geotourism is a form of sustainable tourism that has the potential to serve as a vehicle for promoting local social sustainability. According to the United Nations Educational, Scientific and Cultural Organization (UNESCO), geotourism involves travels through an area, with the tourist being sensitive to the fact that the landscape they are admiring comprises distinct natural features shaped by dynamic processes which have resulted in remarkable land (Sanz et al., 2020). Carrión-Mero et al. (2020) argue that geotourism goes beyond geological tourism; it is a wider notion, the primary interest of which lies in the geological spectacle, the manner in which tourists utilise the natural setting, as well as the cultural traditions and practices of the local populace (Herrera-Franco et al., 2021; Newsome et al., 2022). Saha (2024) notes that the size of the geotourism market worldwide is estimated to reach a value of US$ 1,258.9 billion by the end of 2033. However, the success of geotourism is undermined by limited knowledge and awareness of the influences underlying its development (Matshusa et al., 2020). Geotourism can be seen to be at the confluence of three core elements: forms, processes, and tourism (Fernández et al., 2011). In this regard, forms refer to the natural endowments of landscapes, inclusive of their features and elements, while processes entail the associated tectonic activities, weathering and accumulation, among other things, and tourism comprises the anthropogenic aspect, as evidenced in touristic undertakings and the appreciation of, inter alia, geology and geomorphology (Dowling, 2014; Chylińska, 2019). The success of geotourism is mediated by factors relating to the socio-economic, environmental, and climate conditions of an area (Costa et al., 2020).

Despite the promising role of geotourism in promoting grassroots development, there is a dearth of research on its role as a potent vehicle for local development. The study reported on in the present article investigated the role of geotourism as an instrument for facilitating local development in South Africa. It contributes to the debate on harnessing natural resources to create and stimulate vibrant economies for sustainable development. South Africa is globally renowned for its outstanding cultural, natural, and geological heritage, which makes it an ideal case study.

Methodology

A systematic review was undertaken, which entailed the identification of relevant studies, appraisal of their quality, and summarising of the evidence (Boren & Moxley, 2015). A literature search was done through the Science Direct and Google Scholar platforms. It would have been useful to expand the search to other databases, but it was not possible to examine databases requiring subscription before access. Searches were organised on the basis of Boolean operators (AND, OR, NOT or AND NOT), which were utilised as conjunctions to either merge or exclude keywords during the tailored search, thus generating focused and productive results. Searches included "geotourism and local development", and "geotourism in South Africa". Eligibility criteria for screening were that the article must deal with the topic of geotourism in South Africa, and be related to local development. The search yielded 150 research results related to geotourism. Further scrutiny showed some of the articles to deal with aspects of tourism other than geotourism; these were then excluded, reducing the number of articles to 60. Of the 60 articles remaining, a number dealt with countries other than South Africa, and were therefore excluded. Seven sources were found to relate directly to South Africa; these are summarised in Table 2.1. The reference lists of the selected articles were also

Table 2.1 Core narratives on geotourism in South Africa

Authors	*Title of study*	*Main findings*
Penn-Clarke et al. (2020). Journal of African Earth Sciences.	Geoheritage in the Matjiesrivier Nature Reserve, a World Heritage Site in the Cederberg, South Africa	South Africa's rich biodiversity hotspots such as the Cederberg, coupled with their archaeological findings, offer considerable potential for the development of geotourism for development.
Ruban (2012). Resources Policy	Geoconservation versus legislation and resources policy: New achievements, new questions – Comment on Cairncross (Resources Policy, 2011), The National Heritage Resource Act (1999): Can legislation protect South Africa's rare geoheritage resources?	Effective legislation tailored to the conservation of national geological heritage is needed in protecting geodiversity while optimising the benefits of geotourism.
Reimold et al. (2014). Elsevier	Chapter 3 - Geotourism potential of Southern Africa Editor(s): Ross K. Dowling, David Newsome, Geotourism, Butterworth-Heinemann.	The South African government regards tourism as a tool for developing its economy. This makes it vital to develop tourism niche areas such as geotourism to increase employment and domestic travel.

(*Continued*)

Table 2.1 (Continued)

Authors	*Title of study*	*Main findings*
Matshusa et al. (2020). Acta Commercii	Developing a scale for measuring influential factors towards geotourism development	This study proposes a geotourism development index that highlights the influential factors and activities of different stakeholders contributing towards geotourism development.
Du Preez (2019). Geojournal of Tourism and Geosites	The contribution of geological features to visitor experiences: comparison between two geotourism sites	Geotourism sites host unique, high-quality geological resources, which contribute to enhancing visitors' experiences.
Mukwada & Sekhele (2017). Journal of Asian and African Studies	The potential of community-based geotourism in rural development in South Africa: The case of Witsie Cave Project	South African communities face unique difficulties in harnessing community-based geotourism as an instrument for local economic development. The viability of community-based geotourism projects is mediated by building the local capacity of stakeholders.
Matshusa et al. (2021). Resources	Challenges of geotourism in South Africa: a case study of the Kruger National Park	Geotourism is a novel phenomenon, which has gained traction in South Africa. Nevertheless, its potential contribution to the country's tourism, conservation of natural resources and employment creation require concerted efforts by key stakeholders to harness its true potential.

examined to identify potentially relevant sources dealing with geotourism in South Africa subject to the aforementioned eligibility criteria. The seven articles were then analysed by assessing the text for observable patterns and recurrent themes, condensing the recurrent themes into brief statements and then developing these into paragraphs (Figure 2.1).

Results and discussion

Geotourism is a niche form of sustainable tourism. It lays emphasis on appreciating the earth's geological features in a manner that promotes ecological and cultural awareness and preservation, while being locally useful (Dowling & Newsome, 2010). Moreover, geotourism embraces the creation of geotourism products that conserve geoheritage, nurture communities, create awareness, and uplift

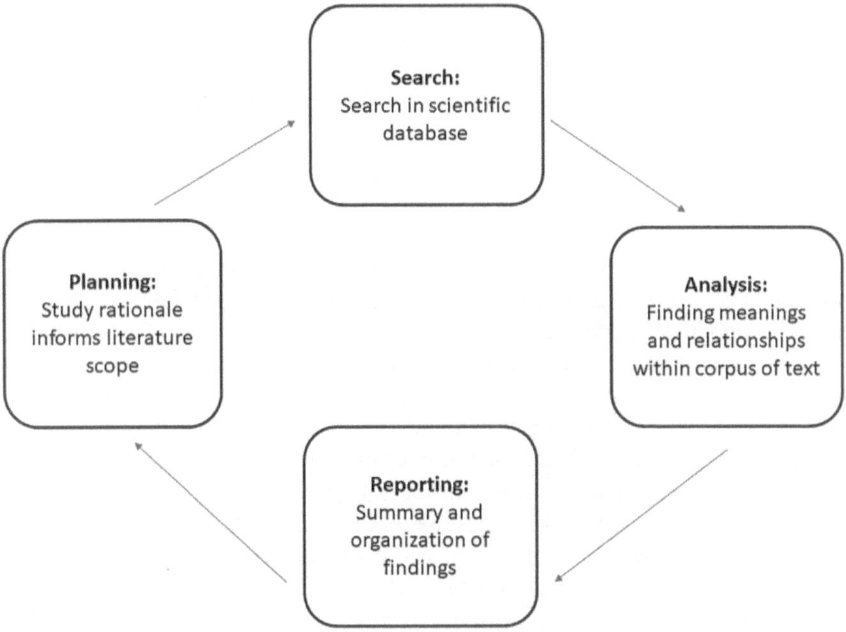

Figure 2.1 Systematic review process conducted during the study.

geological heritage while engaging a cross-spectrum of people (Dowling & Newsome, 2010). The study reported on sought to explore the role of geotourism as a tool for local development.

Promoting World Heritage sites and bio-archaeological hotspots

Penn-Clarke et al. (2020) posit that geotourism is connected primarily to the special identity of a destination. Thus, the UNESCO Global Geoparks (UGGp) has promoted the concept of geoparks as a vital element in geotourism development. Writing almost two decades earlier, Binns & Nel (2002) identified tourism promotion as a tool for local development, and further emphasised the value of geotourism as a vehicle for economic growth, community development, and poverty alleviation in marginalised communities. The promotion of geotourism is accompanied by awareness creation of the unique bio-cultural heritage of destination communities. The harnessing of local environmental resources will help facilitate local economic development, serving as a boost to local government initiatives (Stone & Nyaupane, 2016). The promotion of geotourism will help local economies to thrive and will create opportunities for improved living standards for communities. In addition, the promotion of local geotourism comes with significant advantages. It has been observed that as tourists flock to destination areas, this expands opportunities for local businesses. Ivona (2021) surmises that niche forms of tourism such as geotourism have the potential to greatly transform rural areas and their economies for sustainable local development, particularly since geotourism diversifies the local economy, thereby enhancing the living standards of the local populace. As geotourism is considered a burgeoning field, innovative promotional activities would be helpful in creating awareness of South Africa's unique geotourism resources and the opportunities for engaging communities for enhanced local development (Ngwira, 2015). The World Heritage sites and bio-archaeological hotspots that characterise South Africa's geotourism landscape therefore require suitable publicity as a means to attract geotourists as the target group, with positive ripple effects on local economies for local development.

Rocha & Ferreira da Silva (2014) are also of the view that heritage sites and bio-archaeological hotspots under geotourism provide avenues for rural development. They argue that geotourism can help alleviate unemployment and poverty through the engagement of local communities in geotourism activities and tourism marketing through diverse platforms such as adventure tourism, ecotourism, rural tourism, and health geotourism. It must be borne in mind, however, that there can be no meaningful local development if the health and wellbeing of the populace is not safeguarded. Ridhwan et al. (2022) make the observation that health and prosperity go hand in hand, as healthy communities provide the healthy human capital needed to sustain productivity and growth for sustainable development. In this regard Rocha & Ferreira da Silva (2014) highlight the close connection between geotourism and medical geology, stating that this needs to be explored in realising the potential of geotourism. For example, the consumption of minerals and other related elements for food, water, soil (through geophagy), or dust is observed in many rural communities in Africa. This is often seen during festivities and traditional and religious ceremonies, where the ingestion, inhalation, or dermal absorption of such minerals is common. Moreover, in many rural communities pelotherapy, or "mud therapy", involving the use of mud or clay as a form of therapy, is widespread. These are opportunities to expand the scope of geotourism and its engagement of rural communities for local development.

Providing supporting legislation or a legal framework for the operation and conservation of national geological heritage

Ruban et al. (2022) discuss the legal loopholes in South Africa's tourism sector, especially regarding geotourism, and suggest that these need to be closed. Mutana & Mukwada (2020) argue that the current legal instruments aim at dealing with the most pressing issues in the tourism sector, but that they are fragmented. This has translated into tourism operators working in silos as they cherry-pick policies and use their discretion in terms of the operational guidelines that they adhere to. Given the fragile and often vulnerable state of some geotourism resources, there is a need for synergy in the governance regulations to safeguard them for posterity and promote sustainable tourism.

Mukwada & Sekhele (2017) are of the opinion that overcoming the fragmentation in the legal frameworks will provide a conducive environment in which stakeholders will be able to operate, innovate, and take risks for the expansion of developing fields such as geotourism. This will greatly enhance the viability of community-based geotourism initiatives and the overall role of tourism as a tool for enhancing the standards of living of the poor and marginalised in society. This is even more applicable in post-apartheid South Africa, where the need for community development as a means to alleviate social inequality remains pressing. As proponents of local development emphasise the harnessing of local resources to build local economies, it will not be advisable to leave the fate of such resources in the hands of tourist operators without the requisite legal frameworks to safeguard sustainability and community rights. So, as novel markets develop in the hospitality and tourism sector, the need for effective guidelines becomes increasingly important (Wong et al., 2014). Legislation is important in the tourism sector to ensure optimum functioning and the smooth operation of its diverse stakeholders. As a developing area in the tourism sector, geotourism, with the requisite guidelines, can serve as a pioneer for ethics, equity and conservation of national geological heritage. Issues such as discrimination against people with disabilities in relation to the use of goods, facilities, and services have been reported in the tourism sector (Jee-Eun, 2020) and tourists with disabilities have often had limitations placed on their participation in tourism activities. However, tourism rights under the United Nations Convention on the Rights of Persons with Disabilities aim to remove all discrimination in the tourism sector and make it inclusive. It has become increasingly evident that regulations prohibiting discrimination against persons with disabilities need to be enacted to safeguard the inclusivity of the tourism sector. Institutionalising equity in the tourism sector can be achieved when the various sectors and niches are foregrounded.

Poor marketing of geotourism products, and safety and equity concerns

Awareness about geotourism as a tourism niche in South Africa is low, which undermines its development. This has translated into situations where the needs of interested tourists are not catered for, as they are not assisted in visiting geoheritage sites. As geosites are often located within communities, tourist visits will bring money into local communities as tourists purchase items of interest or allied services from the communities, which will support growth of the local economy. As a way to overcome this, Matshusa et al. (2021) suggest the creation of roles such as geotourism ambassador or champion at geoparks or nature parks, especially those operated by South African National Parks (SANParks), and providing incentives that will make geotourism attractive and easily accessible. This situation also highlights a wider deficiency within SANParks, which needs attention regarding tourism product development and marketing. A number of scholars have underscored the importance of packaging in tourism, as it enhances diversification and specialisation. Poor packaging and marketing of tourism products such as those relating to geotourism can ultimately compromise the appeal of a tourism destination.

This points to the need for a sense of urgency in resolving the problem of poor packaging and marketing in the geotourism space. Closely associated with this is the need for adequate infrastructural support for geotourism development (Martin, 2010). This includes providing ecologically friendly amenities such as roads and toilets in close proximity to target geoheritage sites. It is noteworthy that the provision of such amenities needs to take into consideration the particular difficulties faced by visitors with disabilities. The absence of access roads to geoheritage sites should be rectified by building an environmentally friendly road network as soon as possible (Chakraborty & Jones, 2018; Cooper, 2018). It is also important that the access road infrastructure be built with the needs of tourists with physical disabilities in mind. An important issue relating to geotourism in South Africa that draws considerable attention is that of security and the safety of tourists, as this can be threatened by attacks by wildlife, injury at the hands of poachers, and other threats due to unsafe environments. Dwyer et al. (2009) argue that safety and security are non-negotiable, as they are fundamental to the growth of sustainable tourism activities. If their safety and security cannot be assured, tourists will avoid certain places, thus drastically reducing the tourism visits to that destination. On the other hand, the sacredness and spirituality of geotourism venues needs to be safeguarded in line with the geotourism ideal of observing cultural ethics. This point is considered vital in the use of geotourism as a tool for promoting a positive view of local culture. Equity is a further issue that plagues the local geotourism landscape. Matshusa et al. (2021) highlight the poor access to geotourism opportunities by local community-based businesses owing to entrenched monopolisation by big players in the industry. This calls for requisite interventions to create a level playing field for all industry players.

Conclusion

The quest for social progress is a common thread in all human societies and a bed-rock of civilisation. This has led to diverse approaches towards development. However, in contemporary times, sustainable development has gained policy focus as a form of development that balances economic, social, and environmental imperatives for the good of posterity. One of the core themes in the concept of sustainable development is the essence of development that is led and felt at the grassroots as encapsulated in the SDG 17. The idea of local development hence has gained traction and is echoed in other global policy documents such as the African Union's Vision 2063 (The Africa We Want). A number of avenues for enhancing local development have been foregrounded including geotourism. Geotourism is the form of tourism that exists in response to the unique geographical character of an area, including its environmental conditions, heritage, aesthetics, culture, and the wellbeing of a community. South Africa is host to diverse geological heritage and related resources that make it a potentially important player in the geotourism sector. Moreover, as a rapidly expanding sector, geotourism has the potential to help overcome some of the pressing socio-economic problems affecting the country. This includes the challenge of local development, which is highlighted in the South African National Development Plan (NDP). The extreme poverty in many rural areas has turned these areas into migrant sending areas rather than thriving local economies. A number of South Africa's rich geoheritage sites are situated in rural areas, which makes the case for exploring the role of geotourism as a vehicle for local development. This burgeoning sector requires adequate legal and policy backing to ensure that industry standards are not compromised and that the needs of both customers and the environment are safeguarded. Local businesses will need support to thrive and become pioneers in this rising sector in the tourism industry. Furthermore, safety, security, and ethical issues are everyday problems which need to be solved in order for there to be a conducive operating environment, which the sector needs in order to remain competitive. Ultimately, geotourism holds promise for harnessing resources and endogenous potentialities of a community, to facilitate local development with ripple effects on national and global development.

Key recommendations:

- The World Heritage sites and bio-archaeological hotspots that characterise South Africa's geotourism landscape will need increased publicity so as to bring in more geotourists, with positive impacts on local economies for local development.
- Fields closely related to geotourism, such as medical geology, need increased attention by stakeholders in harnessing their immense potential as a tool for local development. These are also opportunities to expand the scope of geotourism and its engagement of rural communities for local development.
- Legal loopholes in South Africa's tourism sector, including that of geotourism, need urgent attention so as to engender a conducive environment for operators to thrive.
- Furthermore, the fragile and often vulnerable state of some geotourism resources necessitates synergy in the governance regulations to safeguard them for posterity and promote sustainable tourism.
- As a burgeoning field, geotourism presents researchers in South Africa and Africa with a unique opportunity to pioneer research that projects the African voice and policy discourse that will inform global policy.
- Foregrounding issues of inclusivity and equity in the field of geotourism will serve to market it as a niche that has zero tolerance for discrimination and is also disability friendly.
- The sacredness and spirituality of geotourism venues need to be safeguarded in line with the geotourism ideal of observing cultural ethics. This point is considered vital in the use of geotourism as a tool for promoting a positive view of local culture.

References

Adamowicz, M. (2023). Local development model as an element of regional sustainable strategy. *IntechOpen*. https://doi.org/10.5772/intechopen.109534

African Union. (2023). *Agenda 2063: The Africa We Want*. https://au.int/en/agenda2063/overview

Agbenyefia, S. (2022). Innovative financing and local development in Ghana. *Open Access Library Journal*, 9: 1–15. https://doi.org/10.4236/oalib.1109013.

Agbevade, A. (2020). Implementation dynamics of local economic development: Comparative empirical experiences from Ghana's local governance system. *Local Economy*, 35(6): 609–624. https://doi.org/10.1177/0269094220972628

Binns, T., & Nel, E. (2002). Tourism as a local development strategy in South Africa. *The Geographical Journal*, 168(3): 235–247. www.jstor.org/stable/3451338

Boren, S.A., & Moxley, D. (2015). Systematically reviewing the literature: Building the evidence for health care quality. *Journal of Molecular Medicine*, 112(1): 58–62. PMID: 25812277.

Carrión-Mero, P., Borja-Bernal, C., Herrera-Franco, G., Morante-Carballo, F., Jaya-Montalvo, M., Maldonado-Zamora, A., Paz-Salas, N., & Berrezueta, E. (2020/2021). Geosites and geotourism in the local development of communities of the Andes mountains. A case study. *Sustainability*, 13: 4624. https://doi.org/10.3390/su13094624

Chakraborty, A., & Jones, T.E. (2018). Mount Fuji: The volcano, the heritage, and the mountain. In *Natural Heritage of Japan: Geological, Geomorphological, and Ecological Aspects*; Chakraborty, A., Mokudai, K., Cooper, M., Watanabe, M., Chakraborty, S. (Eds). Cham, Switzerland: Springer, pp. 167–175.

Chylińska, D. (2019). The Role of the picturesque in geotourism and iconic geotourist landscapes. *Geoheritage*, 11: 531–543.

Cooper, M. (2018). The lake Aken area: A future geopark? In *Natural Heritage of Japan: Geological, Geomorphological and Ecological Aspects*; Chakraborty, A., Mokudai, K., Cooper, M., Watanabe, M., Chakraborty, S. (Eds). Cham, Switzerland: Springer, pp. 153–159.

Costa, S.P., & Sonaglio, K.E. (2020). Análisis del comportamiento resiliente de los gestores de turismo. *Estudios y perspectivas en turismo*, 29: 331–348.

Diko, S.K., Okyere, S.A., Opoku Mensah, S., et al. (2021). Are local development plans mainstreaming climate-smart agriculture? A mixed-content analysis of medium-term development plans in semi-arid Ghana. *Socio-Ecological Practice Research*, 3: 185–206. https://doi.org/10.1007/s42532-021-00079-2

Doan, K.H. (2022). The role of community for local development: A research agenda. *The USV Annals of Economics and Public Administration, Stefan cel Mare University of Suceava, Romania, Faculty of Economics and Public Administration*, 22(1(35)): 7–17.

Dowling, R.K. (2014). Global geotourism – an emerging form of sustainable tourism. *Czech Journal of Tourism*, 2.

Dowling, R.K., & Newsome, D. (2010). Global geotourism perspectives. *KIP Articles*, 2262. https://digitalcommons.usf.edu/kip_articles/2262

Du Preez, E.A. (2019). The contribution of geological features to visitor experiences: Comparison between two geotourism sites. *GeoJournal of Tourism and Geosites*, 26(3): 874–888. https://doi.org/10.30892/gtg.26316-404

Dwyer, L., Edwards, D., Mistilis, N., Roman, C., & Scott, N. (2009). Destination and enterprise management for a tourism future. *Tourism Management*, 30: 63–74.

Fernández, M.P., Timón, D.L., & Marín, R.G. (2011). El geoturismo como estrategia de desarrollo en áreas rurales deprimidas: Propuesta de geoparque villuercas, Ibores, Jara (Extremadura). *Boletin de la Asociacion de Geografos Espanoles*: 485–497.

Fourie, W. (2018). *Aligning South Africa's National Development Plan with the 2030 Agenda's Sustainable Development Goals: Guidelines from the Policy Coherence for Development movement*. Pretoria: University of Pretoria.

Herrera-Franco, G., Montalván-Burbano, N., Carrión-Mero, P., Jaya-Montalvo, M., & Gurumendi-Noriega, M. (2021). Worldwide research on geoparks through bibliometric analysis. *Sustainability*, 13: 1175.

Ivona, A. (2021). Sustainability of rural tourism and promotion of local development. *Sustainability*, 13(16): 8854. https://doi.org/10.3390/su13168854

Jee-Eun, L. (2020). Understanding the discrimination experienced by customers with disabilities in the tourism and hospitality industry: The case of Seoul in South Korea. *Sustainability*, 12(18): 7328. https://doi.org/10.3390/su12187328

Kisman, Z.A., & Tasar, I. (2014). The key elements of local development. *Procedia Economics and Finance*, 15: 1689–1696. https://doi.org/10.1016/S2212-5671(14)00642-X

Martin, S. (2010). The Grand Canyon, USA: The experience of managing a world-class geotourism destination. In *Global Geotourism Perspectives*; Dowling, R.K., Newsome, D. (Eds). Oxford, UK: Goodfellow Publishers, pp. 167–178.

Matshusa, K., Leonard, L., & Thomas, P. (2021). Challenges of geotourism in South Africa: A case study of the Kruger national park. *Resources*, 10: 108. https://doi.org/10.3390/resources10110108

Matshusa, K., Thomas, P., & Leonard, L. (2020). Developing a scale for measuring influential factors towards geotourism development. *Acta Commercii*, 20(1): a861. https://doi.org/10.4102/ac.v20i1.861

Mensah, J.K., Bawole, J.N., & Ahenkan, A. (2017). Local economic development in Ghana: From the 'lost decades' to a policy 'maturing' stage. *Development Southern Africa*, 34(5): 607–621. https://doi.org/10.1080/0376835X.2017.1310032

Mukwada, G., & Sekhele, N. (2017). The potential of community-based geotourism in rural development in South Africa: The case of Witsie cave project. *Journal of Asian and African Studies*, 52(4): 471–483. https://doi.org/10.1177/0021909615595991

Mutana, S., & Mukwada, G. (2020). Are policies and guidelines shaping tourism sustainability in South Africa? Critical success factors for tourism sustainability governance in the Drakensberg region. *Tourism and Hospitality Research*, 20(2): 198–209. https://doi.org/10.1177/1467358419841100

Newsome, D., Ladd, P., & Dowling, R. (2022). The scope for geotourism based on regolith in Southwestern Australia – a theoretical and practical perspective. *Geoheritage*, 14(5): 55–62. https://doi.org/10.1007/s12371-021-00632-1

Ngwira, P.M. (2015). Geotourism and geoparks: Africa's current prospects for sustainable rural development and poverty alleviation. In *From Geoheritage to Geoparks, Case Studies from Africa and Beyond*; Errami, E., Brocx, M., Semeniuk, V. (Eds). Cham, Switzerland: Springer.

Penn-Clarke, C.R., Deacon, J., Wiltshire, N., Browning, C., & du Plessis, R. (2020). Geoheritage in the Matjiesrivier nature reserve, a world heritage site in the Cederberg, South Africa. *Journal of African Earth Sciences*, 166(2020): 103818. https://doi.org/10.1016/j.jafrearsci.2020.103818

Reimold, W.U., Whitfield, G., & Wallmach, T. (2014). Geotourism potential of Southern Africa. In *Geotourism*; Dowling, R.K., Newsome, D. (Eds). London, United Kingdom: Routledge.

Ridhwan, M.M., Nijkamp, P., Ismail, A., et al. (2022). The effect of health on economic growth: A meta-regression analysis. *Empirical Economics*, 63: 3211–3251. https://doi.org/10.1007/s00181-022-02226-4

Rocha, F., & Ferreira da Silva, E. (2014). Geotourism, medical geology and local development: Cape Verde case study. *Journal of African Earth Sciences*, 99(2): 735–742. https://doi.org/10.1016/j.jafrearsci.2014.04.015

Rogerson, C.M., & Rogerson, J.M. (2010). Local economic development in Africa: Global context and research directions. *Development Southern Africa*, 27(4): 465–480. https://doi.org/10.1080/0376835X.2010.508577

Ruban, D.A. (2012). Geoconservation versus legislation and resources policy: New achievements, new questions – Comment on Cairncross (Resources Policy, 2011), The National Heritage Resource Act (1999): Can legislation protect South Africa's rare geoheritage resources? *Resources Policy*, 37(1): 126–129. https://doi.org/10.1016/j.resourpol.2011.12.005

Ruban, D.A., Mikhailenko, A.V., & Yashalova, N.N. (2022). Valuable geoheritage resources: Potential versus exploitation. *Resources Policy*, 77(2022): 102665. https://doi.org/10.1016/j.jafrearsci.2020.103818

Saha, S. (2024). *Geotourism Market Outlook (2023–2033)*. https://www.futuremarketinsights.com/reports/geo-tourism-market#:~:text=The%20global%20geotourism%20market%20size,5.5%25%20from%202023%20to%202033 accessed on 10.07.2023

Sanz, J., Zamalloa, T., Maguregi, G., et al. (2020). Educational potential assessment of geodiversity sites: A proposal and a case study in the Basque Country (Spain). *Geoheritage*, 12(23): 20–45. https://doi.org/10.1007/s12371-020-00432-z

Stone, L.S., & Nyaupane, G.P. (2016). Africans and protected areas: North – South perspectives. *Annals of Tourism Research*, 58: 140–155.

UNDP. (2023). *The SDGs in Action*. www.undp.org/sustainable-development-goals accessed on 04.03.2024

Wong, G., Derthick, A.O., David, E.J.R., Saw, A., & Okazaki, S. (2014). The what, the why, and the how: A review of racial microaggressions research in psychology. *Race and Social Problems*, 6: 181–200.

3 Geotourism as a contribution to social and ecological justice

Contemporary approaches in sub-Saharan Africa

Michael Dyssel

A social-ecological analysis of geoheritage resource uses and livelihoods concerns of villagers in the Cederberg Mountains, Western Cape, South Africa

Introduction

The importance of the abiotic, geological, and geomorphological endowment of environments is often overshadowed by their biotic and biodiversity attributes (Gray, 2003). This apparent bias is changing though if the establishment of geoparks worldwide is anything to go by. China, with its "289 national geoparks and 41 UNESCO global geoparks" (Xu & Wu, 2022: 1) is arguably the most important beneficiary of geotourism and its economic and environmental spin-offs. Gordon (2020) believes that geotourism and geoconservation are important drivers for more holistic protected area and conservation management. Understanding the natural origins and evolution of geoheritage resources – whether landscapes, landforms, archaeological, and palaeontological sites or processes – is key to successful geoconservation and geotourism practices, specifically if the latter impacts communities and their livelihoods (Ólafsdóttir & Tverijonaite, 2018). Geodiversity refers to the variety and attributes of the geological and geomorphological features that characterise specific locations and regions. Georesources relate to geodiversity attributes of an area to which value is attached and from which economic benefits can be accrued. Geoconservation deals with the conservation of geological and geomorphological features, largely based on preservationist approaches to sensitive geodiversity areas such as palaeontological and fossil sites.

South Africa has rich natural and cultural diversity that positions the country as a premier tourist destination. The country's biodiversity, varied landscapes and apex big-five species (i.e. lion, elephant, buffalo, leopard and rhino) are huge drawcards for domestic and international tourism. The popularity of the country's biotic endowment often overshadows its abiotic natural endowment. Knight et al. (2015: 165) therefore believe that "the geologic and geomorphological diversity of South Africa, the spectacular landscapes that result, and their internationally important ecological and cultural/archaeological associations, all make the landscapes of South Africa prime sites for geoheritage and geotourism". Subsistence and poor communities in

DOI: 10.4324/9781032706764-3

rural settings are often marginalised from the tourism benefits that can be derived from the bio-, geo-, and cultural-historical attributes that shape their life worlds and livelihoods. The exploitation of such attributes by market-driven tourism is well documented worldwide and is associated with critical notions such as the 'staging' of, 'gazing' at, and 'othering' of people in their traditional, in-situ locations. Matshusa et al. (2021) also highlight the important role that geotourism can play in the improvement of people's livelihoods as it can – directly and indirectly – contribute to general Sustainable Development Goals (SDGs), specifically SDG 1 that deals with poverty alleviation and SDG 8 that deals with employment creation. Access by and association of local people with tourism resources and its benefits are crucial for maintaining sustainable tourism and conservation management in disenfranchised locations (Dowie, 2009; Hayward, 2016; Brockington & Wilkie, 2015; Dyssel, 2020). In this respect it is important and useful to distinguish between geo-related concepts such as geodiversity, georesources and geoconservation.

The Cederberg forms the western leg of the Cape Folding Belts and associated mountain ranges in the Western Cape, developed through orogenic (mountain building) processes in the Triassic period, 300 million to 230 million years ago (Quick & Eckardt, 2015; Compton, 2015). The extensive sandstone and shale deposits of the Cape Supergroup were subsequently subjected to faulting and folding to create a rugged landscape characterised by ridges and high mountain peaks, such as Sneeuberg, more than 2000 m above the sea level. Mountain ridges are separated by valleys drained by perennial streams such as the Tra-Tra, Sand, Eselbank, Matjiesfontein, and Biedouw Rivers, as well as non-perennial streams in and adjacent to the Cederberg Wilderness Area (CWA) (see Figure 2.1). The geology and geomorphology of the area, in combination with the Mediterranean climate of the region, shape conditions conducive to the area's Fynbos (largely in the southern-central parts) and the Succulent Karoo vegetation (towards the northern parts). Quick & Eckardt (2015) think that it is of vital importance that "the interlinked geology, geomorphology and ecology are protected as part of the Cederberg Wilderness Area, which is a significant geotourism and geoheritage region, rich in archaeological remains".

The Cederberg is indeed one of South Africa's more pristine wilderness areas with very unique biodiversity, geodiversty, and cultural-historical diversity endowment (Wilson, 2015; McIntosh, 2017; Van Schalkwyk, 2019; Faurie, 2020). The historical and current livelihood practices of economically deprived villagers found in proximity to this declared wilderness complex are challenging. (Van Schalkwyk, 2019). Settlements such as Wupperthal, Heuningvlei, Eselbank, and Langkloof are located in different sections of the Cederberg (Figure 3.1). The ancient footprint of First Peoples, who inhabited this mountainous area and of which villagers are descendants, such as the San (hunter-gatherers) and Khoikhoi (pastoralists), is reflected at various rock art sites in the Cederberg. Rock art sites also reveal that a different ecology and habitat must have existed, as captured by well-preserved sketches of eland, elephant and leopard that used to roam the area. The Cederberg's geology and geomorphology are characterised by some of the most spectacular landscapes and landforms, largely comprised of Table Mountain Sandstone, outcropping in high mountain peaks such as Sneeuberg, Sneeukop, and Tafelberg. The Wolfberg Arch,

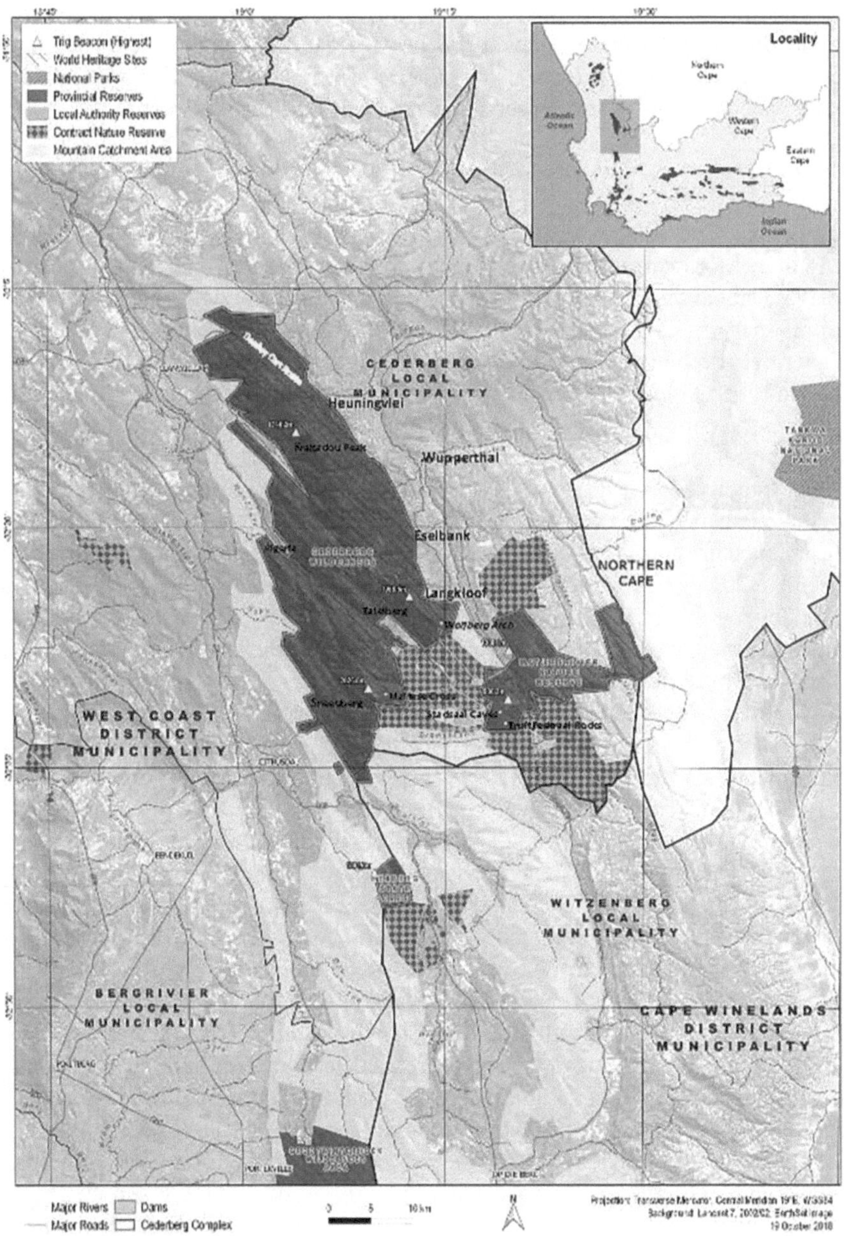

Figure 3.1 Location of settlements such as Wupperthal, Heuningvlei, Eselbank, and Langkloof in the Cederberg.

the Stadsaal Caves, and the Maltese Cross are impressive and highly visited signature landforms in the area. Coupled with a moderate Mediterranean climate, these geodiversity attractions present unique nature-based and ecotourism opportunities that private landowners and conservation authorities are cashing in on.

Case study areas

Wupperthal: The village, located at 32° 16′ 27″ S, 19° 12′ 33″ E, was established in 1830 by Moravian missionaries of German origin. It is surrounded by 14 outpost villages, including Heuningvlei in the north and Eselbank and Langkloof in the central-southern parts, on the eastern edge of the CWA. It is surrounded by homoclinal ridges dipping in a general north-north-easterly direction and bisected by the Tra-Tra River. Franklin & Breed (2017: 41) believe that Wupperthal "is one of the best-preserved mission stations in the Western Cape that still functions as an active community under the administration of the Church". This, along with the settlement's largely subsistence livelihood practices in a remote and generally harsh environment, makes it a very special place (Franklin & Breed, 2017; Heyns, 1980). The architecture and social-cultural fabric of the village was impacted negatively at the end of 2018 when most of the traditional white-washed thatched-roofed houses burnt down. Most of these houses, which date back to the 1830s when the village was established, were devastated by a huge fire that left the larger part of the village in ruins and most inhabitants homeless. The rebuilding of the village is taking place at a very slow place Begg (2021). Other economic activities associated with Wupperthal such as the Rooibos tea industry, the *veldskoen* shoe factory and the tannery that provides leather to the shoe factory were negatively affected by the fire.

Heuningvlei: The village, located at 32° 12′ 12″ S, 19° 05′ 58″ E, is one of the outposts of Wupperthal and is the most important cultural and religious hub near the CWA. The small settlement consists of approximately 25 families residing in largely thatched-roofed cottages. The original Heuningvlei Farm belonged to several brothers, with the surname Ockhuis, who inherited the property from their father. Just as in the case of Wupperthal, most inhabitants are involved in subsistence agriculture and livestock practices. (Rememberers n.d.). Traditional methods to produce dried fruit and cultivate Rooibos tea organically are also practised by the inhabitants of Heuningvlei. Tourism-related activities in Heuningvlei include backpacker and community guest-lodge accommodation and the donkey cart route between the Pakhuis Pass and Heuningvlei (Destination Cederberg, n.d.). The Pakhuis Pass is a scenic part of Route 364 between the towns of Clanwillian and Calvinia; other attractions include hiking routes, off-road 4×4 and jeep track routes, guided walks with locals during the wildflower season (July–September) and guided rock-art talks by locals on the donkey cart route (Tracks4Africa, n.d.).

Heuningvlei will benefit from a bioprospecting agreement between CapeNature and the The Body Shop International company, as the village plays host to a project that re-introduces the endemic and critically threatened "cedar tree (*Widdringtonia cedarbergensis*) into its natural environment in the Cederberg". (CapeNature is the provincial conservation authority in the Western Cape). The cedar tree is critically endangered and is included on the red list of threatened species of the International Union for the Conservation of Nature. The purpose of the project is for The Body Shop International "to access scent samples from the leaves and dead wood of the cedar tree – described as 'biological resources' – to produce its Widdringtonia Eau De Parfum". It is argued that the project will be fundamental to enhancing the ecotourism potential of the area, recognising the cultural and heritage value of the cedar tree and improving the general livelihoods of people in the village (CapeNature, n.d.).

Eselbank: The village is located at 32° 21′ 01″ S, 19° 13′ 29″ E and its name is related to the Afrikaans word 'esel' which means donkey. Just as is the case in Wupperthal and Heuningvlei, villagers are largely subsistence-oriented. They are mainly involved in the rooibos tea and livestock farming in, and adjacent to, the floodplain of the Eselbank River that runs through the village. Others, mainly women, are involved in the production and selling of handicrafts and home-made products such as bread and different jams.

Langkloof: The village is located at 32° 24′ 41″ S; 19° 14′ 42″ E on the south-eastern edge of the CWA. It is the smallest of the four villages under investigation and is in essence characterised by the same (subsistence) livelihood conditions as the others. It is closest to some of the most popular geotourism sites such as the Wolfberg Arch, Maltese Cross, Sneeuberg, and Tafelberg in the CWA, as well as the Stadsaal Caves and Truitjieskraal Rock Formations in the Matjiesrivier Nature Reserve.

Literature review and theoretical aspects

This study fits into a general sustainable livelihood framework which recognises that impacts of context-determined natural, social, cultural, and economic conditions and processes can shape the well-being of people. However, an appreciation for the complex interactions and interdependencies between ecological and social-development concerns is needed before any sustainable solutions for social-ecological challenges can be devised. Virapongse et al. (2016: 85) reckon that "SES science frames relationships between human and ecological components as part of a complex system with multi-scale feedbacks and dependencies" Ecological components are often the essence of conservation efforts that, in turn, are facilitated by the establishment of protected areas. The increase in the spatial extent and the number of protected areas, managed by organs of state or through partnerships and stewardship arrangements, has a variety of direct and indirect impacts on local people.

It is now recognised worldwide that such impacts need to be understood before their consequences can be addressed sustainably (Brockington & Igoe, 2006; West et al., 2006; Büscher & Dressler, 2007). As resource restrictions for locals are often associated with the establishment and extension of protected areas, the latter often bears the brunt of discontent, as reflected in anti-conservation activities and transgressions such as poaching, illegal harvesting, and land invasion. It is therefore crucial that instead of pursuing blanket preservationist approaches to protected area management, sustainable benefit sharing of resources and ecological services that can contribute to local people's livelihoods should be promoted. Ecological and social systems are therefore inseparable. Wilson (2015) therefore, believes that "conservation approaches that aim to achieve more resilient systems must take these social systems into consideration". The remoteness, relatively untouched landscapes, and general inhospitable environment of the Cederberg Mountain and the related proclaimed wilderness area created a sparsely populated 'frontier area' for those residing there. Historically, it shaped the spatiality and temporalities of livelihood practices in the mentioned villages since the occupation of the area. As explained, such practices are largely related to small-scale agricultural production on the floodplains and the valleys of the Tra-Tra River near Wupperthal, the Biedou River near Heuninglvei, and the Eselbank River near Langkloof and Eselbank. Agricultural activities include the production of vegetables, deciduous fruit, and rooibos tea. These practices are augmented by livestock rearing that primarily includes sheep, goats, pigs, and chickens. (Commercial agricultural practices on larger tracks of land owned primarily by white farmers, are more dispersed and patched across the Cederberg.) Environmental influences, such as cyclical drought conditions, along with the general lack of other economic activities in these villages expose the vulnerabilities of the people who are dependent on subsistence activities (Franklin & Breed, 2017; Heyns, 1980). It is therefore important that the basis for addressing livelihoods under precarious conditions needs to be recognised and extended to include benefit sharing from natural-resource endowment such as geo-, bio-, and cultural-historical diversity in the study area (Van Schalkwyk, 2019). As discussed in the preceding sections, conservation initiatives in this regard have been established in Heuningvlei's donkey cart drives from Pakhuis Pass, the cedar tree initiative, guided tours, etc.

The shift in development-conservation management from centralised to more cooperative and inclusive practices that recognise the role and place of local people in South Africa is non-negotiable for sustainable conservation (Chevallier, 2016; Wilson, 2015). Sustainable conservation in resource-scarce areas can only be achieved through the collective and communal use of natural resources, whether actively consumptive (i.e. extractive) and/or actively non-consumptive (i.e. through responsible tourism). This falls within the ambit of "community-based natural resource management (CBNRM) [that] has had mixed results since the first programmes were initiated in Southern Africa" (Chevallier, 2016: 2). Despite the critique against CBNRM, variations on the approach can yield positive circumstance-specific results. The CWA was managed by the Department of Forestry prior to 1994, but is now under the jurisdiction of CapeNature, the official conservation agency in the Western Cape province. CapeNature's jurisdiction includes stewardship and conservancy agreements with private landowners. Although local folks' resource uses were restricted since 1890, villagers seemingly had a broader resource base from which they could eke out a living, specifically people of Heuningvlei, Eselbank, and Langkloof. (Personal conversations, 2021 & 2022; Van Schalkwyk, 2019; Wilson, 2015). They could sustain their livelihoods by using natural resources such as buchu, wild rooibos tea, and firewood and by using former farmland to sustain their subsistence livestock practices.

The restrictions imposed by CapeNature are enshrined in Acts, and their subsequent amendments, such as the National Environmental Management Act (NEMA), Act 107 of 1998 and its related sectoral NEMAs such as the National Environmental Protected Areas Act (NEMPAA), Act 57 of 2003; the National Environmental Biodiversity Act (NEMBA), Act 10 of 2004; and the National Heritage Resources Act (NHRA), Act 25 of 1999. Importantly, Van Schalkwyk (2019) believes that the CWA is increasingly becoming a haven for eco- and nature-based tourism and that long-established rural communities practising subsistence farming in the Cederberg need to get a fairer share of ecotourism. Some communities reside on the very boundary of the CWA, causing some degree of land use disconnects relative to tourism and conservation initiatives. Tourism and conservation "are currently reframing the Cederberg as a leisure landscape" (Van Schalkwyk, 2019: vii; Wilson, 2015). The fact that the area is becoming a primary leisure landscape is not always compatible with livelihood practices, such as livestock rearing and small-scale agricultural production of local inhabitants. Tourism is often used as leverage to address both conservation and livelihood concerns in proximity to protected areas and to unlock further potential in such areas. Significantly, Faurie (2020: iii) warns that, if not practised responsibly and with strategic-social accountability, tourism activities could cause protected areas to "become self-destructive".

Methods and tools

In the context of the social-ecological systems (SES) approach, a survey method was followed for this study and it comprised of data-gathering tools such as observations, site visits, and conversations with locals. A conversational approach, which is less formal than interviews, underpinned the data gathering process that took place (over four weekends) in September 2021 and March 2022. Typical snowball sampling was used, as the villages were relatively far from each other and poor road quality and difficult terrains needed to be traversed and negotiated. It was difficult to set up meetings in advance as telecommunications (even at the best of times) was challenging, specifically in Heuningvlei, Eselbank, and Langkloof. With the assistance of a local guide with good knowledge of the CWA and the villages, older people were targeted and approached for conversations. Notably, the older folk could relate to periods when restrictions on resource uses for livelihood purposes in the Cederberg were less restrictive. The concepts and impacts of geodiversity, georesources, and geoheritage had to be explained in more detail to some participants, who had limited formal schooling and understanding. Conversations went smoothly though, as all the participants as well as the local guide were Afrikaans-speaking. Although permission was granted by some participants to have conversations audio-recorded, as long as they were treated anonymously, field and memory notes of conversations were kept by the researcher. (The latter was acutely aware that the participants, despite their plight, had agency that needed to be respected.)

As Wupperthal is the biggest educational, cultural, and religious hub of the area, it was easier to get people who wanted to participate in the study; therefore, more conversations (10) could be 'snowballed' there. There was a general willingness, during both fieldwork periods, by locals to participate in conversations and visits to sites that they considered to be of geo- and cultural-heritage value. The general reluctance by most participants to be audio recorded was respected, as per research ethics practices, and written fieldnotes were compiled only after conversations. Two participants in Wupperthal permitted fossils and striking rock fragments that they had collected over the years to be photographed (Figures 2(a), (b) and (c)). Conversations and questions interrogated the following aspects: demographic variables and profiles of participants, popular geotourism sites of the CWA and other protected areas, village-specific geoheritage resources, and the latter's (potential) role in the improvement of livelihoods. The content of collected information was analysed according to themes that included previous and current resource uses, the level of sustainability of such uses, the enabling and/or disenabling livelihood circumstances caused by the CWA, and the promotion of local geoheritage options to diversify livelihoods.

Data presentation and analysis

Table 3.1 is a summary of the number of conversations and profiles of each of the participants in the different villages. Although snowball sampling underpinned the survey, attempts were deliberately made, with the help of a local guide, to gain insight from diverse people in the different villages (as is evident in the table).

Table 3.1 Conversations and profile of participants

	Heuningvlei	*Wupperthal*	*Eselbank*	*Langkloof*
Number of conversations	5	10	4	3
Profile of participants	Subsistent crop farmer (2) Handyman (1) Accommodation manager (1)	Retired teacher (2) Clergyman (1) Subsistent crop farmer (3) Livestock farmer (2) Shop assistant (1) University student (1)	Handicrafter (1) Unemployed youth (1) Subsistent crop and livestock farmer (2)	Subsistent crop and livestock farmer (3)

Although villagers recognise the importance of popular, market-driven geotourism attractions such as the Wolfberg Arch, Maltese Cross, Stadsaal Caves, and Truitjieskraal Rock Art site and their role in geotourism in the southern parts of the CWA, participants expressed the sentiment that many locals felt alienated from the mentioned attractions. (It is largely owing to accessibility and the fact that most visitors, due to poor road quality, do not traverse the landscape further north than the Matjiesrivier Nature Reserve, as four-wheel drive vehicles with a high road clearance are needed. Locals who need to go to neighbouring, bigger towns such as Clanwilliam and Citrusdal do it in a northerly direction via Wupperthal.) Table 3.2 summarises the geoheritage sites/resources that participants felt were of cultural value to them and that they did not necessarily want to be commercialised for livelihood purposes. The value attached to these locations should be appreciated against the background of decades of intergenerational subsistence as well as religious practices. Visits were paid to the locals' geoheritage sites, accompanied by some of the respondents and the local guide, to the locations listed in Table 3.2.

The visuals in the following sections show some geoheritage resources that could be promoted more in support of local livelihoods instead of a dependency on the market-driven geotourism sites from which some locals feel alienated.

Although the owners and collectors of the fossils and rock fragments could not provide the exact locations of where the specimens were found, they could provide the broader areas in which it was found. Figures 3.2(a), (b), and (c) show a selection of such specimens.

Figure 3.2(a) shows two almost identical pear-shaped fragments that were reported to have been picked up, lying side by side, on an extremely hot day near a

Table 3.2 Local geoheritage sites and popular geotourism sites

	Heuningvlei	*Wupperthal*	*Eselbank*	*Langkloof*
Local/ Community geoheritage sites	Caves Rock art sites	Singkop Spitskop Kroeke Dam Tra-Tra River * Vaalheuning * Die Bakke * Hartsvlak	Waterfall Rooibos fields Sneeuberg	Waterfall Rock art sites Ceder bush Khoisan graves Tafelberg
	* Some high-lying areas outside Wupperthal where activities such as livestock grazing and rooibos tea fields (used to) occur			
Popular and market-driven geotourism sites of the CWA	Wolfberg Arch; Wolfberg Cracks; Maltese Cross; Lot's Wife and Window Rocks; Stadsaal Caves and Rock Art sites; Sneeuberg, Tafelberg; plenty of hiking trails, bouldering locations; mountain bike trails and 4×4 trails; abseiling locations			

rock embankment on the plateau in the Vaalheuning region east-southeast of Wupperthal. The participant believed, rightly or wrongly so, that the fragments bear testimony of a dinosaur era that Wupperthal needs to be recognised for.

Figure 3.2(b) shows the *Tentaculites* Brachiopod (CapeNature, 2019: 57) embedded in a sandstone conglomerate. Based on the classification of Penn-Clarke et al. (2018), the specimen resembles carbonate nodules of the Gydo Formation. The owner mentioned that it was picked up somewhere near Eselbank and that many similar specimens could be found all over the area. The participant had a good understanding of the origin of the Cederberg Mountains and their dominant peaks (Sneeuberg and Tafelberg) and reckoned that efforts should be made to popularise the village of Eselbank more by also focusing on its geoheritage endowment.

Figure 3.2(c) resembles what seems to be an *Australospirifer* Brachiopod (CapeNature, 2019: 59) that are largely restricted to the Bokkeveld Group sediments that outcrop significantly in the Matjiesrivier Nature Reserve towards the south of the CWA. The specimen was found near the Brugkraal Settlement en route to Heuningvlei, according to the participant. The participant understood the formation of fossils and did not express any opinion as to how such finds could contribute to the betterment of local people.

Figure 3.2(a) Two almost identical pear-shaped fragments.

Figure 3.2(b) Shows the *Tentaculites* Brachiopod.

Figure 3.2(c) Resembles an *Australospirifer* Brachiopod.

Figures 3.3(a), (b), (c), and (d) show some geoheritage sites that also have particular cultural significance for locals in the different villages. Although villagers believe that such sites can be promoted to generate some income, they are not necessarily in favour of mass visitations to such sites.

Figure 3.3 (a) Rock stacks near Langkloof and (b) Eselbank Waterfall.

Source: Ina's Paradise Guesthouse (n.d.). *Source:* Ina's Paradise Guesthouse (n.d.).

Figure 3.3 (c) Spitskop near Wupperthal (d) Rock art near Heuningvlei.

Source: Wupperthal trots, 2014

Source: Cedarberg Afrika (n.d.).

Figures 3.4(a), (b), (c) and (d) show some popular geotourism sites in the CWA for which permission and payment are required before they can be visited. The Wolfberg Cracks (i.e. deep and narrow thoroughfares through the massive Table Mountain sandstone strata that lead to the spectacular Wolfberg Arch) are

Figure 3.4 (a) Wolfberg Cracks and (b) Wolfberg Arch.

Source: Slingsby Maps 2013 Source: SafariNow (n.d.)

arguably the most popular geotourism sites in the area. Similar are the impressive Maltese Cross, shaped by mechanical weathering over millions of years, and the Stadsaal Caves. The latter, excavated and shaped by weathering and erosion agents, got its name from gatherings held there by former apartheid-era prime ministers and cabinet members. For example, the name of D.F. Malan (former prime minister) is still graffitied on the cave walls.

Figure 3.4 (c) Maltese Cross and (d) The Stadsaal Caves.

Source: Cedarberg Africa 2023

Analysis of conversations

As mentioned, a conversational approach, which is less formal than interviews, underpinned the data-gathering process and ties in with SES, CBNRM and deep tourism as theoretical lenses for sense making. SES, by virtue of its specific focus on social and ecological mutuality that underpins complex systems, is an appropriate lens to reflect on the CWA and the communities that were studied. The spatial proximity of the study areas to the CWA, literally on its eastern edge, makes an appreciation of the social and ecological mutuality discussed even more important. As participants understood the importance of the CWA as an almost site-aside area to be preserved for ecological purposes, they needed to ensure the preservation of their livelihoods in that context as well. As highlighted, prior to the declaration of the CWA as a World Heritage Site, administered by Cape Nature, the framework for the active consumptive resource-use practices between the communities and the CWA could be considered as CBNRM. Participants considered that relationship as largely sustainable, as they – just as their forefathers – use natural resources, such as patches of arable and grazing land and water, for subsistence and not commercial purposes.

As the CWA has become more popular as a tourist destination, it is important that local communities be included in the sharing of benefits accrued from the commercialised biodiversity, geodiversity and cultural attractions of the areas. Importantly, this sentiment was expressed by most of the participants. Likewise, ecotourism was specifically mentioned, by those with a tertiary education, as a vehicle to unlock further potential and sharing of benefits. It can therefore be argued then that deep ecotourism, as a specific form of ecotourism, be further promoted to include the geoheritage potential and resources of the communities. As an extension of the geoheritage potential of the area, an appreciation of the religious activities practised by people of Wupperthal on the Table Mountain sandstone ridge named Singkop, south of the village, is a further case in point. Annually, during the Easter period, people in the village undertake a 'pilgrimage' up the ridge to sing traditional Moravian hymns and to acknowledge and commemorate the death of Jesus Christ. Participants in this study felt that such practices should never be commercialised and that outsiders and visitors, albeit free to join such activities, always need to be respectful. Fennell & Singh (2004:109) consider deep ecotourism as a vehicle for the incorporation of the "hard and soft path-forms of ecotourism", providing a basis for the incorporation of the values and ethics of people associated with ecotourism destinations.

Although respondents of Langkloof, in particular, understood the importance of the conservation of the highly threatened endemic and ceder tree, they deplored the fact that access to ceder wood was restricted to local people who had used it sustainably for decades. The manufacturing and selling of handicrafts and wood-carving art were mentioned as ways in which locals could add to their livelihoods. Although they did not cut down growing ceder trees themselves, they used the roots and branches to create art after the straight stems of the trees were used primarily by the then Post Office as support poles for its land-based telecommunications network. Dry ceder wood collected in the area was used as fuel in the past but due to its recognised conservation value, crafty locals also used it for the manufacturing of furniture and wooden cutlery. The mountain passes and thoroughfares of the Cederberg provide motorised as well as non-motorised access to the area's geoheritage sites, the villages of the case studies and smaller hamlets in proximity to the villages.

Ross (2011) believes that South Africa's mountain passes and thoroughfares are, over and above their historical engineering underpinnings, also testaments to the challenges and tenacities of peoples that they connect. Iconic mountain passes in the study area such as the Pakhuis, Krakadou, and Koueberg Passes (in the north), the Niewoudts and Uitkyk Passes (in the south), and the Old Postal Route (towards the east, linking the Twanka Karoo and the Cederberg) were also negotiated by villagers on foot for decades. Older participants nostalgically reminisced about traversing the area in their quests to sustain their livelihoods through livestock herding, rooibos tea harvesting, wood collection, and labouring on commercial farms. Participants also appreciated the fact that improved livelihood opportunities rooted in the geoheritage options would always be subject to the temporalities of other contributing attractions such as heavy snowfall and the natural wildflower display in the area during South Africa's winter and early spring seasons.

Overall, participants in the villages were unanimous in their appreciation of any new initiatives that would contribute to their livelihoods by promoting and developing their own geoheritage and cultural resources in sustainable ways. Younger and more vociferous respondents alluded to the fact their existence on the edge of the CWA needs to be taken into consideration in attempts to diversify livelihoods. The Heuningvlei Donkie Cart Route was heralded as an example of good synergy between conservation tourism and employment of local people. Their knowledge of the biodiversity (specifically the mountain fynbos species), landforms, and rock art contributed significantly to the success of the route. The uniqueness of the study, therefore, lies in the opinions of locals on how to sustainably benefit from commercialised and non-commercialised geoheritage sites in and around an ecologically-sensitive wilderness area.

Conclusion

The investigation in the CWA yielded interesting spatial-economic dynamics associated with (local) geoheritage resource uses in the study area. Inhabitants of the villages that border the protected areas and who are largely dependent on intergenerational subsistence practices do have very specific challenges related to conservation restrictions. The study highlights the complexities of the conservation–livelihoods nexus in and in proximity to a World Heritage Site (i.e. the CWA), formal protected areas (nature reserves), and contractual conservation areas managed by private landowners. Although tourism is fundamental to the success of conservation areas and can serve as a tool for managing conservation–livelihood challenges, it can become – by virtue of how it is practised – detrimental to conservation mandates. Holistic area-based environmental governance and sensitive location-based conservation management of and in proximity of protected areas are therefore important to support the different forms of tourism on offer in an area like the Cederberg.

Deep ecotourism can, for example, support the advantages that may be accrued from the locals' own geoheritage sites in combination with the commercialised geotourism sites that are located mainly in the south of the CWA. Although inseparable, locals think that access to resources associated with abiotic features such as landforms and landscapes need to be less restricted compared to access to biodiversity resources – based on a sustainable livelihoods narrative. The spatial footprint of South Africa's conservation estate needs to be extended based on its rich conservation endowment. However, it needs to be done with the necessary holism and sensitivity. The consolidation and expansion of protected and conservation areas are guided by legislative prescripts that need to be followed (NEMA, 1998; NEMPAA, 2003; NPAES, 2008, 2010, 2016; NPAESRD, 2009; NHRA, 1999). Although CapeNature works according to a so-called strategic implementation framework that takes cognisance of the social, biological, economic, cultural, and institutional challenges of the Cederberg Complex as a protected area, it will be essential to develop and facilitate long-term sustainable participatory synergies and action plans with local people who have deep roots and bring their own inherent 'value propositions and contributions' to the future success of the CWA and its surroundings.

Within the context of the general initiatives of CapeNature to marry conservation and communities' livelihood concerns across its different protected areas in general, and the challenges of the villages on the edge of the CWA in particular, the following recommendations would be worth (re)considering:

- Local people's indigenous knowledge of the natural environment needs to be promoted and facilitated more effectively for conservation and tourism sustainability in the area.
- The rich geoheritage of the Cederberg and the CWA, as relatively pristine locations in the Western Cape, can indeed become spaces for exemplary convivial conservation by strengthening stewardship relations.
- Capacity building and opportunities need to be extended to involve more local people in the commercialised geotourism-related offerings in the Cederberg.
- The love and affinity of the people of Wupperthal, Heuningvlei, Eselbank, and Langkloof for the Cederberg and their intergenerational loyalty towards their religion (Moravian) and subsistence lifestyles are – and will continue to be – strong pillars for the promotion ecological and social justice.

References

Begg, A. (2021). *Historic Wupperthal rebuilds after 2018 fire*. Southern and Eastern African Tourism Update, 15 November. Online: www.tourismupdate.co.za/article/historic-wupperthal-rebuilds-after-2018-fire (Accessed 18 August 2022)

Brockington, D. and Igoe, J. (2006). Evictions for conservation: A global overview. *Conservation and Society* 4(3): 424–470.

Brockington, D. and Wilkie, D. (2015). Protected areas and poverty. *Philosophical Transactions of the Royal Society B* 370(1681). https://doi.org/10.1098/rstb.2014.0271

Büscher, B. and Dressler, W. (2007). Linking neoprotectionism and environmental governance: On the rapidly increasing tension between actors in the environment-development nexus. *Conservation and Society* 5(4): 586–611.

CapeNature. (2019). *Cederberg complex: Protected area management plan 2019–2029*. Cape Town: Internal Report CapeNature.

CapeNature. (n.d.). *Heuningvlei community and cedar trees benefit from Body Shop perfume agreement*. Online: www.capenature.co.za/projects/heuningvlei-community-and-bioprospecting-on-cedar-trees (Accessed 2 September 2023)

Cedarberg Africa. (2023). *Walks in the central Cederberg wilderness*, 18 December. Online: www.cedarberg-travel.com/walks-central-cederberg-wilderness-reserve (Accessed 23 February 2023)

Cedarberg Afrika. (n.d.). *Klein Krakadouw trail*. Online: www.cedarberg-travel.com/holidays/klein-krakadouw-trail (Accessed 8 April 2022)

Chevallier, R. (2016). *The state of community-based natural resource management in Southern Africa: Assessing progress and looking ahead. Occasional paper 40*. Johannesburg: South African Institute of International Affairs.

Compton, J.S. (2015). *West coast – a natural history*. Cape Town: Earthspun.

DestinationCederberg. (n.d.). *Heuningvlei hike*. Online: https://thewinetender.co.za/tour-heuningvlei.html (Accessed 3 October 2021)

Dowie, M. (2009). *Conservation refugees: The hundred-year conflict between global conservation and native peoples*. Cambridge, MA: MIT Press.

Dyssel, M. (2020). Conservation tourism challenges and opportunities on the Cape Flats, South Africa. In: L. Leonard, R. Musavengane & P. Siakwah (eds.), *Sustainable urban tourism in sub-Saharan Africa – risk and resilience*, pp. 196–215. Abingdon-on-Thames: Routledge.

Faurie, M. (2020). A critical analysis of ecotourism and good environmental governance in the Cederberg Complex. Master's thesis. University of Stellenbosch.

Fennell, D.A. and Singh, T.V. (2004). Deep ecotourism: Seeking theoretical and practical reverence. In: T.V. Singh (ed.), *New horizons in tourism: Strange experiences and stranger practices*, pp. 109–120. Wallingford: CABI Publishing. https://doi.org/10.1079/9780851998633.0109

Franklin, M. and Breed, I. (2017). Wupperthal, an object of people and place. *South African Journal of Art History (SAJAH)* 32(2): 39–55.

Gordon, J.E. (2020). Geoconservation principles and protected area management. *International Journal of Geoheritage and Parks* 7(4): 199–210. https://doi.org/10.1016/j.ijgeop.2019.12.005

Gray, M. (2003). *Geodiversity: Valuing and conserving abiotic nature*. London: Wiley and Sons.

Hayward, N. (2016). *Protected area management effectiveness: Facilitating conservation management [lecture]. Biodiversity review*. Cape Town: CapeNature.

Heyns, H. (1980). *Wupperthal 150 jaar, 1830–1980 – Evangeliese Broederkerk in Suid-Afrika*. Genadendal: Genadendalse Drukkery.

Ina's Paradise Guesthouse. (n.d.). *Photo and video gallery*. Online: http://inasparadise-guesthouse.co.za/main/the-area-and-activities/ (Accessed 3 October 2021)

Knight, J., Grab, S. and Esterhuysen, A.B. (2015). Geoheritage and geotourism in South Africa. In: S. Grab & J. Knight (eds.), *Landscapes and landforms of South Africa: World geomorphological landscapes*. Cham: Springer International Publishing. https://doi.org/10.1007/978-3-319-03560-4

Matshusa, K., Leonard, L. and Thomas, P. (2021). The contribution of geotourism to social sustainability: Missed opportunity? *The International Journal of Sustainability in Economic, Social and Cultural Context* 17(1): 95–118. https://doi.org/10.18848/2325-1115/CGP/v17i01/95-118

McIntosh, F. (2017). The Cederberg rocks. *Country Life magazine*, March.

National Environmental Management Act (NEMA) No. 107 of 1998. (1998). Government Gazette 401(19519). Online: www.gov.za/sites/www.gov.za/files/a107-98.pdf

National Environmental Management: Protected Areas Act (NEMPAA) No. 57 of 2003. (2003). Government Gazette 464(26025). Pretoria: Government Printers.

National Heritage Resources Act (NHRA) No. 25 of 1999. (1999). Government Gazette 406(19974). Online: www.gov.za/documents/national-heritage-resources-act

National Protected Areas Expansion Strategy for South Africa (NPAES). (2008). *Priorities for expanding the protected area network for ecological sustainability and climate change adaptation*. Pretoria: Government Printers.

National Protected Area Expansion Strategy for South Africa (NPAES). (2010). *Department of environmental affairs*. Pretoria: Government Printers.

National Protected Areas Expansion Strategy for South Africa (NPAES). (2016). *Department of environmental affairs*. Online: www.dffe.gov.za/sites/default/files/docs/national_protectedareas_expansionstrategy2016_ofsouthafrica.pdf

National Protected Area Expansion Strategy Resource Document (NPAESRD). (2009). *Department of environmental affairs and South African national biodiversity institute*. Online: www.environment.gov.za/sites/default/files/docs/npaes_resource_document.pdf

Ólafsdóttir, R. and Tverijonaite, E. (2018). Geotourism: A systematic literature review. *Geosciences* 8: 234. https://doi.org/10.3390/geosciences8070234

Penn-Clarke, C.R., Rubidge, B.S. and Jinnah, Z.A. (2018). High-paleolatitude environmental change during the early to middle Devonian: Insights from Emsian – Eifelian (lower – middle Devonian) siliciclastic depositional systems of the Ceres subgroup (Bokkeveld group) of South Africa. *Journal of Sedimentary Research* (88): 1040–1075. https://doi.org/10.2110/jsr.2018.53

Personal conversations with participants, September 2021 and March 2022.

Quick, L. and Eckardt, F. (2015). The Cederberg: A rugged sandstone topography. In: S. Grab & J. Knight (eds.), *Landscapes and landforms of South Africa*, pp. 85–93. New York: Springer.

Rememberers. (n.d.). *The centre for curating the archive*. Online: www.rememberers.co.za/people/johannes-ockhuis/ (Accessed 8 September 2021)

Ross, G. (2011). *The romance of cape mountain passes*. Cape Town: Sunbird Publishers.

SafariNow. (n.d.). *Wolfberg cracks and arch*. Online: www.safarinow.com/destinations/cederberg-wilderness- area/hikingtrail/wolfberg-cracks-and-arch.aspx (Accessed 4 March 2021)

Slingsby Maps. (2013). *Place names in the Cederberg*, 28 July. Online: https://cederberg-names.blogspot.com (Accessed 4 March 2021)

Tracks4Africa. (n.d.). *Heuningvlei*. Online: https://tracks4africa.co.za/listings/item/w128362/heuningvlei/ (Accessed 5 June 2023)

Van Schalkwyk, I. (2019). Accommodating the wild? Leopard conservation, tourism and local communities in the Cederberg. MA thesis. University of the Western Cape, Bellville.

Virapongse, A., Brooks, S., Metcalf, E.C., Zedalis, M., Gosz, J., Kliskey, A. and Lessa, L. (2016). A social-ecological systems approach for environmental management. *Journal for Environment Management* 178: 83–91.

West, P., Igoe, J. and Brockington, D. (2006). Parks and peoples: The social impact of protected areas. *Annual Review of Anthropology* 35: 251–277.

Wilson, J.M.R. (2015). The benefits and burdens of living beside the Cederberg wilderness area. MSc thesis. University of Cape Town.

Wupperthal Trots. (2014). 28 September. Online: https://wupperthal.blogspot.com/2014/09/beskrywende-skryfstuk_82.html (Accessed 13 December 2023)

Xu, K. and Wu, W. (2022). Geoparks and geotourism in China: A sustainable approach to geoheritage conservation and local development – a review. *Land* 11(9): 1493. https://doi.org/10.3390/land11091493

4 Impacts of climate change on geoheritage and geotourism

A Southern Africa Perspective

Khodani Matshusa

Introduction

Climate change has emerged as one of the major threats to the sustainability of geoheritage sites and geotourism. The world is facing a growing climate crisis (Hill & Kimball, 2023). The concept of climate change has received a lot of attention and evolved over time. However, the impacts of climate change on geoheritage and geotourism have received little attention, especially in the Global South (Matshusa et al., 2021a; Mehdioui et al., 2022) and in particular in Southern Africa (Dube & Nhamo, 2018). Geoheritage, geotourism, and geoparks are expanding rapidly in the Global North, with the Global South (especially Southern Africa) lagging behind. Geoheritage refers to the elements of the Earth that we value, such as landforms, rocks, soils, minerals, fossils, and active geological processes (e.g. volcanic activity) (Newsome & Dowling, 2018). Geoheritage sites have become important resources for achieving sustainable development through geotourism (Ehsan et al., 2013; Matshusa & Leonard, 2023; Mehdioui et al., 2022). Geotourism can be summarised as sustainable tourism that is based on the existence of geoheritage sites (Štrba et al., 2023). Geoparks are "single, unified geographical areas where sites and landscapes of international geological significance are managed with a holistic concept of protection, education and, sustainable development" (UNESCO, 2018). In this context, the 2023 the United Nations Educational, Scientific and Cultural Organization (UNESCO) report highlights how climate disruption was affecting communities and ecosystems across UNESCO Global Geoparks (UGGPs) and Biosphere Reserves in Latin America and the Caribbean (UNESCO, 2023). However, there are few similar initiatives in the Global South, especially Southern Africa. This chapter adds to prior studies (Dube & Nhamo, 2018; Dube & Nhamo, 2020; Chikodzi et al., 2022) by highlighting the impacts of climate change on geoheritage and geotourism with a focus on Southern Africa.

The natural capital in the Southern Africa region is a major contributor to economic growth and fiscal revenue, driving investment in physical and social infrastructure. However, the region is particularly at risk of the devastating impacts of human activities and climate change on the natural capital. As such, the region is facing two main challenges: (i) lack of capacity to mitigate and build resilience to climate change and (ii) lack of strategies for protecting the natural capital and

DOI: 10.4324/9781032706764-4

ensuring human well-being, which must be tackled together (African Development Bank, 2023). To address these challenges, Aspiration 5 of the African Union's first 10-year implementation plan (2013–2023) identifies heritage as one of the priority areas of Agenda 2063 "The Africa We Want". Furthermore, Priority 5 of the Southern African Development Community (SADC) Regional Indicative Strategic Development Plan (RISDP) 2020–2030 and SADC Vision 2050 seeks to promote sustainable environment and natural resources management (African Development Bank, 2023). As such, tourism initiatives like the utilisation of geotourism, geoheritage, and geoparks have the potential to stimulate investment and sustained socioeconomic growth while contributing to geoconservation, cultural identity, and climate action (Matshusa et al., 2021b). Geoconservation is the conservation of geoheritage sites for their intrinsic, ecological, and geologic or scientific values (Sharples, 2002).

As of October 2023, the Southern Africa region had a total of 44 World Heritage Sites (WHS), with South Africa (10 WHS) having the highest number of sites on the continent. Although the region has remarkable and diverse geological and mineral wealth, geoheritage, geotourism, and geoparks remain largely unknown in Africa (Matshusa & Leonard, 2023). To date, there is no regional structure or institution responsible and advocating for the preservation of remarkable geoheritage sites in Southern Africa. Despite the global growth of geoconservation, geoheritage, and geotourism (Quesada-Valverde & Quesada-Román, 2023; UNESCO, 2023), there is limited research (Dube & Nhamo, 2018, 2020; Chikodzi et al., 2022; Matshusa & Leonard, 2023) on the impacts of climate change on geoheritage and geotourism in the Southern Africa region. Furthermore, understanding perceptions and reactions to the impacts of climate change will be essential to future geoheritage and geotourism developments. However, to understand and manage the impacts of climate change on geoheritage sites requires collaborative research involving the geotourism research community, destination managers, and operators (Gordon, 2023). Considering this limited research, this study presents qualitative research on impacts of climate change on geoheritage and geotourism. In addition, this study presents research results from three iconic African protected sites – the Cradle of Humankind World Heritage Site (CoH-WHS), the Magaliesberg Biosphere Reserve (MBR), and the Kruger National Park (KNP) – as a benchmark for Southern Africa, if not Africa. These sites were selected due to their geological and human-cultural history (CoH-WHS and MBR) and the interlinkages between geology, culture and wildlife (KNP).

Literature Review

Climate change impacts on geoheritage: an overview

Geoheritage and geotourism contribute to social sustainability (Matshusa et al., 2021b) and local economic development. To understand the value and potential of geoheritage sites and geotourism to sustainable tourism, it is vital to identify the ecological impacts, threats, or risks to the sites from climate change events and natural hazards such as floods. Climate change poses serious threats to the scientific integrity and cultural and aesthetic values of geoheritage sites globally. Therefore, management actions in response to climate change impacts on geoheritage sites should factor in the geological, cultural and aesthetic values (Reimold, 1999; Crofts et al., 2020; Gordon et al., 2022). Moreover, climate-related events are worsening, which will further undermine geoheritage potential for socioeconomic development regionwide (Mc Keever & Narbonne, 2021). Similar threats have been identified in the United Kingdom (Cigna et al., 2018).

Southern Africa is projected to experience serious climate change impacts such as intense rainfall, droughts and excessive heat waves (Engelbrecht et al., 2024). Gauteng province is projected to experience unprecedented impacts such as day-zero drought (Engelbrecht et al., 2024), while Limpopo province (especially the KNP) will have high temperatures, heavy rainfall, and the possibility of flooding as key climatic threats (Chikodzi et al., 2022). Research has identified human-induced and nature-induced impacts on geoheritage and geotourism (Matshusa & Leonard, 2023). Man-made events pose threats to geoheritage sites through activities such as visitation (overcrowding) by tourists, deforestation, construction, mining, theft, and vandalism. These human activities can exacerbate nature-induced impacts on geoheritage sites and geotourism. Incinerators contribute to climate change by sending out enormous amounts of carbon dioxide into the atmosphere and pollutants into the environment, thereby heightening climate change impacts on natural heritage (Carmona & Gil, 2022; Morriss, 2022). According to Zhang et al. (2022), anthropogenic factors pose a much higher threat to natural heritage than natural factors, in terms of both type and number. In this context, degradation of geological sites is mainly due to human negligence and lack of knowledge, which leads to public misuse and mismanagement of the geoheritage (Selmi et al., 2022).

Natural processes such as floods (Crofts et al., 2020), rock falls, vegetation encroachment, and erosion (El Hassi & Muftah, 2022) can damage geoheritage and geotourism sites. Geoheritage sites are prone to climatic processes such as high temperatures, heavy rainfall and floods (El Hassi & Muftah, 2022). Gordon et al. (2022) provide an overview of direct and indirect potential physical impacts on geoheritage, including features and sites with value for geotourism. Indirect impacts on geoheritage include human activities such as deforestation and disruption of geomorphological processes. Direct impacts on geoheritage are due to geomorphological processes (progressive melting of glaciers and permafrost, sea-level rise, and increased frequency and severity of extreme weather events, including flash floods, droughts, storm surges at the coast, hurricanes, and tropical storms) (Gordon, 2023). For nature-induced or direct impacts, it is common that geological features tend to be significantly damaged by catastrophic events rather than ongoing daily, seasonal, and annual routine processes (Woo & Worboys, 2019). Drastic climatic conditions such as high temperature, humidity, and the amount of precipitation control the chemical weathering of rock outcrops (Woo & Worboys, 2019), including geological sites. Geoheritage sites close to the coast are more vulnerable to water and wind processes (Selmi et al., 2022). For example, extreme weather events are predicted to negatively impact the Maltese Islands (Selmi et al., 2022).

Extreme water levels and floods can affect the stability of the ground and result in collapse of geoheritage sites (Woo & Worboys, 2019). Climate change by extended drought poses serious and persistent threats to Lake Ohrid on the border between the Republic of Macedonia and the Republic of Albania (Carmona & Gil, 2022; Chalovska et al., 2022).

Overview of climate change and geoheritage: legislative framework

The protection of geoheritage from natural hazards and climate change is still not being prioritised by international organisations such as the United Nations (Reynard & Brilha, 2018). Furthermore, the Earth Summit in Rio de Janeiro and Agenda 21, Sustainable Development Goals, and the World Tourism Organisation have left the importance of geoheritage, geotourism, and geoconservation out of consequential international discussions. By recognising geoheritage and geotourism as sustainable tourism components, the UNESCO declaration on the protection of geological sites of international significance and geoparks highlighted the establishment of the regional geoheritage and geoparks committees with the mandate to monitor and address the impacts of climate change on natural sites such as geoheritage sites. To monitor the progress made in managing climate change events in regions, UNESCO, the International Union of Geological Sciences (IUGS), and the International Union for the Conservation of Nature (IUCN) identified the need to study the impacts of climate change on geoheritage by member states. The IUGS and UNESCO established the International Geosciences and Geoparks Program (IGGP), which culminated in the formation of the UNESCO Global Geoparks Network (UGGN) in 2001.

The African Geoparks Network is a member of the UGGN. In May 2024, out of 213 UGGPs, there were only two geoparks in Africa (the M'Goun geopark in Morocco and the Ngorongoro-Lengai geopark in Tanzania) and none in Southern Africa. During the Conference of the Parties to the United Nations Framework Convention on Climate Change 26, UNESCO highlighted the contribution of UGGPs throughout their territories to climate action and adaptation. The IUCN identified climate change as the most common threat to natural WHS listed under criterion viii (geology), such as geoheritage sites and geoparks (Osipova et al., 2020). As such, the IUCN Geoheritage Specialist Group formally established the Geoconservation and Climate Change Working Group (GCCWG) in 2022, with the main task of geoheritage conservation for the future. According to Hill & Kimball (2023), the goal of the GCCWG is to promote nature-based solutions for effective adaptations and solutions to mitigate the impacts of climate change. These institutions have been advocating for the preservation of geoheritage sites at a global and continental levels.

The IUCN's Guidelines for Geoconservation in Protected and Conserved Areas provide best practices for managing the impacts of climate change on geoheritage and geotourism (Crofts et al., 2020). Best Practice Guideline 18 requires a multi-step approach to address threats to geoheritage, including identifying the types of threats, the sensitivity of the site to the threat, risk assessment and prioritisation of management actions (Crofts et al., 2021). Furthermore, Gordon (2023) provides a global overview showing that the consequences of climate change have been recognised in geotourism research. According to Gordon (2023), there are limitations in the existing studies on climate change regarding geoheritage and geotourism. This suggests a lack of capacity due limited expertise and academics to conduct research on the impacts of the climate on geoheritage and geotourism. Furthermore, there is lack of legal frameworks and policy drivers relating to geoheritage and geotourism in most countries (Prosper et al., 2018).

In the context of Southern Africa commitments through the Copenhagen Accord, the Paris Agreement and the Glasgow Climate Pact, advanced economies promised financial support for climate assistance to developing countries, including countries in Southern Africa. However, these promises have not been met. Sub-Saharan Africa (especially Southern Africa) is affected by increasingly intense and frequent natural disasters related to climate change (IMF, 2022). The region is particularly at risk of the devastating impacts of human activities and climate change on the natural capital. As mentioned previously, the region is facing three main challenges: (i) mitigating and building resilience to climate change; (ii) protecting the natural capital; and (iii) ensuring human well-being, which must be tackled together (African Development Bank, 2023). To address these challenges, aspiration 5 of the African Union's first 10-year implementation plan (2013–2023) identifies heritage as one of the priority areas of Agenda 2063 "The Africa We Want". Furthermore, Priority 5 of the SADC RISDP 2020–2030 and SADC Vision 2050 seek to promote sustainable environment and natural resource management. As such, tourism initiatives such as the utilisation of geotourism, geoheritage, and geoparks have the potential to stimulate investment and sustained socioeconomic growth while contributing to geoconservation, cultural identity and climate action (Matshusa et al., 2021a; African Development Bank, 2023). Furthermore, studies on adaptation and streamlining government policy will be key to geoheritage (Aligishiev et al., 2022; Bellon & Massetti, 2022). For the region, more concessional climate commitments on adaptation and geoconservation will be critical to supporting geotourism development. Geoconservation is a broad concept of conservation, protection, and management of geoheritage (Schluter & Schumann, 2018).

Currently, there is no regional legal framework and institutional capacity to specifically manage the impacts of climate change on geoheritage sites in Southern Africa. Without proper administration and proper regulations (Quesada-Valverde & Quesada-Román, 2023) to manage the impacts of climate change and associated threats, geoheritage and geotourism in Southern Africa are at risk. Managing climate change impacts will assist in minimising or preventing the degradation of geoheritage sites. The impact of climate change on geoheritage varies depending on geographic locations. For example, geoheritage sites in very warm temperatures are associated with physical and chemical weathering, while geoheritage sites in cold or rainy environments are prone to floods. Southern Africa is known for its unique WHS and protected areas with international geological and cultural significance such as the CoH-WHS, MBR, and KNP. In addition, the region is globally known for its rich biodiversity, wildlife, culture and geology, especially in protected areas. South Africa is home to one of the largest protected areas in Africa – the KNP, where various geoheritage sites were identified for geotourism development (Matshusa et al., 2021a, 2021b). Despite the rich geological history, southern Africa currently has no plans and legislation to manage climate change impacts on geoheritage sites (Matshusa & Leonard, 2023). Consequently, geoconservation in protected areas is crucial for protecting, sustaining, and managing geoheritage sites and biodiversity from climate change impacts such as floods and erosion (Crofts, 2019). There is a need to engage the public and communities alongside other stakeholders in wildlife and cultural heritage (Prosper et al., 2018; Chikodzi et al., 2022). Research shows a lack of engagement of key stakeholders to inform adaptation planning and decision making by geoparks, geotourism destination managers and other stakeholders (Gordon, 2023; Kubalíková et al., 2023). To fill this gap, the methodology of this study involved stakeholder engagement with key informants through semi-structured interviews with a focus on the CoH-WHS, MBR (widely known for human and cultural evolution) and KNP (popularly known for its wildlife diversity).

Methodology

Study area

The selected study sites for the research were the CoH-WHS, MBR, and KNP. The CoH-WHS is a UNESCO World Heritage Site with excellent geological areas, such as the Sterkfontein Caves known for historic fossil finds, and the MBR is a UNESCO Biosphere Reserve rich in natural, cultural, and human assets. These include the magnificent Magaliesberg mountain range. The KNP contains important geoheritage and cultural heritage areas such as the Makahane (sandstone cave and dinosaur fossils), Mashikiri (sandstone arc, cave, and portholes), and Thulamela (sandstone walling) geoheritage sites with geological heritage and geotourism potential, as described by Matshusa et al. (2021a, 2021b). The three study sites were selected because they contain World Heritage with both geological and human-cultural history (CoH-WHS and MBR) and represent the interlinkages between geology, culture, and wildlife (the KNP). Natural attractions such as the CoH-WHS, MBR, and KNP are rated as top attractions in Southern Africa (Matshusa & Leonard, 2023). These sites are some of the main sites (a national park, a world heritage site, and a world biosphere reserve) with geoheritage and geotourism potential in southern Africa.

The COH-WHS, in north-west Gauteng, was experiencing an economic downturn in local agriculture and mining activities when the decision was taken in May 1997 to nominate it for WHS recognition (Rogerson & Van der Merwe, 2016).

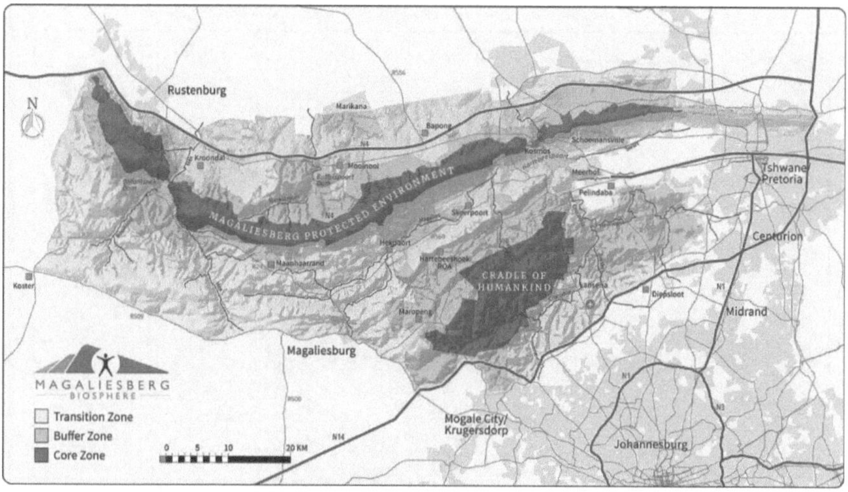

Figure 4.1 Location of the CoH-WHS as part of the Magaliesburg Biosphere Reserve.
Source: Magaliesberg Biosphere Reserve, 2021

The site area is about 40 km from Johannesburg and borders Mogale City to the south, the City of Tshwane to the north-east, and the North West province to the east (Figure 4.1). It is viewed as a potential lever to regenerate the local economy of this economically depressed region through heritage tourism (Rogerson & Van der Merwe, 2016). From a climatological point of view, the CoH-WHS is characterised by warm subhumid temperature with about 759 mm of precipitation annually. Rainfall predominately occurs during summer, with most precipitation falling in January (averaging 142 mm); in winter, there is much less rainfall than in summer (Shapi et al., 2021). The MBR comprises the CoH-WHS and extends for about 357,870 ha from Pretoria in Gauteng province to Rustenburg in the North West province (Figure 4.1).

The KNP is in Limpopo province, at the north-eastern corner of South Africa (Figure 4.2). It is the largest national park in South Africa and one of the largest national parks in Africa, one of the top tourist destinations in the world, and borders the countries of Zimbabwe to the north and Mozambique to the east. It contains some of the important relics of ancient civilisation, trade and development of the Southern Africa region. The KNP has high temperatures, heavy rainfall and the possibility of flooding (Chikodzi et al., 2022). The CoH-WHS, MBR, and KNP are surrounded by some of the poorest and marginal communities of Gauteng, North West, and Limpopo provinces. This highlights the need to create opportunities for the local communities to develop businesses using natural resources such as the geoheritage sites in the CoH-WHS, MBR, and KNP through geotourism.

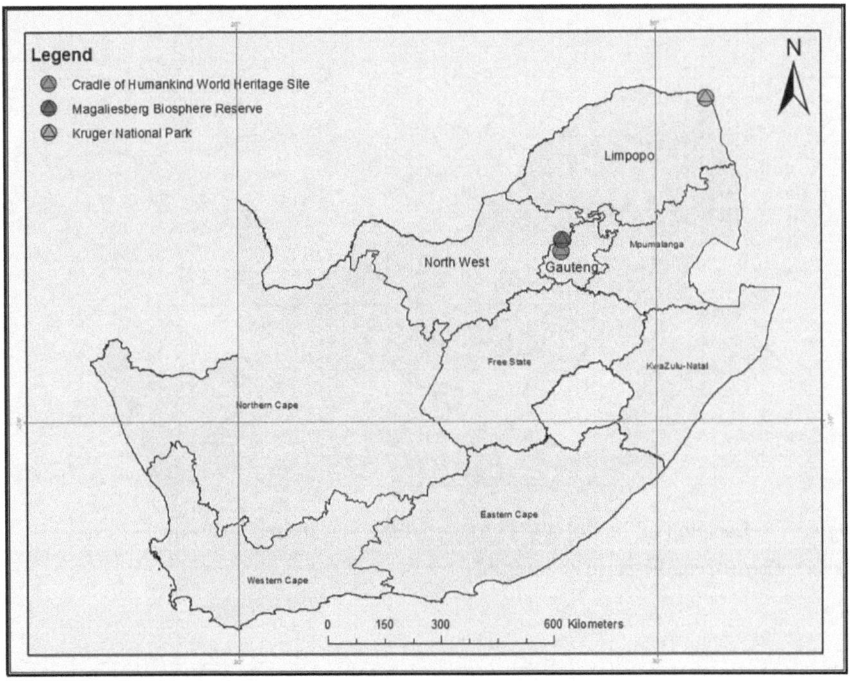

Figure 4.2 Location of the study sites.

Methods

The study adopted a qualitative approach to conduct semi-structured interviews with key stakeholders from the selected sites, local communities, the government, and the private sector. The study endeavoured to understand how the impacts of climate change were perceived by different key stakeholders. The questions that were asked focused on understanding the impacts or threats of climate change on geoheritage and geotourism. During the data collection, semi-structured interviews were used to collect data from the key stakeholders (Table 4.1). This gave all the key stakeholders an opportunity to provide the researcher with answers on issues concerning the impacts of climate change on geoheritage sites and geotourism. The interviews were conducted online and, in some cases, one on one with stakeholders in venues selected by the participants.

The purposive sample size was based on the interviewees' unique knowledge and/or understanding of the topic and the availability of the key informants to take part in this research. As such, semi-structured interviews were conducted with a total of 30 key participants. The participants from the CoH-WHS and MBR were clustered into one group because the sites are geographically linked (the CoH-WHS

Table 4.1 List of stakeholders interviewed

Participants	*Participant codes*	*Number of participants*	
		KNP	*CoH-WHS and MBR*
Senior officials at SANParks/KNP and CoH-WHS and MBR	SANParks: AVJ, JS, KM, TK CoH-WHS/MBR: X, L	4	2
Senior government officials	TM, BM, MB, MP, PM, NS	6	
Tourism agencies/forums/ operators	SANParks: MM1, MM2, AY CoH-WHS/MBR: G, W, R, H, A	3	5
Education institutions	SANParks: NĹC, MBR CoH-WHS/MBR: T, B	2	2
Local community leaders	SANParks: R, TS CoH-WHS/MBR: MBT	2	1
Local residents	SANParks: PM, TI CoH-WHS/MBR: HG	2	1

site is within the MBR). All the senior South African National Parks (SANParks), CoH-WHS, MBR, and government officials who participated were recruited by direct visits to the study areas and offices. The representatives of education institutions, local leaders, and residents were recruited through a mediated process by SANParks, CoH-WHS, and MBR officials.

The local community leaders and residents at the KNP were selected after consultation and recommendation by the Mutele Traditional Authority Council. Tour operators within the CoH-WHS and MBR were recruited through referrals from officials during interviews, while an online approach was used to identify and recruit tour operators at the KNP. It is worth noting that during the interviews, some participants suggested other potential informants. Each interview lasted approximately 60 minutes. The interviews were electronically recorded and then transcribed; however, some participants did not consent to be recorded. Content analysis (Creswell, 2014) was used to generate meaningful themes from the participants' responses. The validity and reliability of the data were ensured by verifying the interview transcripts against the interview recordings.

Results and discussions

Climate impacts on geoheritage sites at the CoH-WHS and MBR

Geographically, the CoH-WHS is part of the MBR within the same climate region, hence the need to discuss the results together. Most of the participants who were interviewed indicated that climate change has the potential to damage geoheritage sites and limit geotourism development. Climate change events such as global warming, flooding, pollution, weathering, and erosion may damage the integrity of geoheritage sites and geotourism.

As noted by NS, a senior official in the national Department of Tourism:

> Climate change is a serious matter. So, tourism often it is about the natural attractions. . . . Climate change across all its facets and its impact is negative. . . . Whether there are water resources, whether they topographic elements such as mountains or valleys . . . landscapes, flora and fauna. . . . All these things are adversely impacted by climate change. So, from a fauna point of view, animals will start breathing differently, different cycles and then some will go into extinction and therefore will make the whole food chain change completely. So, if I want to take you to [the] Magaliesberg biosphere to see, let's say, [a] particular rabbit, you will find that that rabbit is no more there. . . . Therefore, all of that impacts on tourism is just that, not everybody sees it the same way. But for me, the impact will always be adverse. . . . And in addition, there might be some gases that are triggered by either flooding or climate change. So then when you want to visit a fossil site such as the CoH-WHS or Sterkfontein underground, but then if you have such gases because of climate change, you can't go there. You are not allowed to go because you can't breathe. So yes, to your point, it [climate change] will have an adverse effect

This view by NS confirms the findings by Durand et al. (2010), who warned that flooding, mining, and pollution present a significant risk to the biodiversity, tourism, and sustainability of the CoH-WHS and MBR. Field observation and interview findings show that geoheritage sites in the CoH-WHS and MBR were being damaged by erosion on high and low slopes (Figure 4.3).

In addition, participants from the CoH-WHS and MBR indicated that the impacts of climate change also caused infrastructure to collapse, resulting in secondary impacts such as pollution and sewage spilling into rivers and other water bodies. There is a risk of acid mine drainage affecting the caves and archaeological sites at the CoH-WHS (Shapi et al., 2021). Participants suggested that the infrastructure of the CoH-WHS and MBR was degraded due to poor maintenance, which necessitated reliable infrastructure to support geotourism. The results indicate that flooding could damage properties and cause loss of revenue. In worst-case scenarios, these climate change activities might result in the tourism establishment closing temporarily or permanently. Informant G from one of the private reserves in the CoH-WHS and MBR noted:

> Global warming, there is a lot of negative implications you know on that. As government infrastructure continues to collapse, you see sewage being dumped in the rivers. Through rain, you know things are washing away, not being able to repair it. So, it's [climate change] impacted our rivers in our

Figure 4.3 The eroded cliffs of Magaliesberg.
Source: Author, 2024

> area quite considerably . . . as seasons come and go, you can see that things do start getting affected. Your patterns change in terms of the weather and the rain. . . . I have the Crocodile River running through my property. I can see what it does to the life of the fish, and it could be very challenging and continued to be more challenging as resources diminish. . . . Certainly, we lost revenue in November 2022. We encountered quite a lot of damage on our property. We didn't have to close. The level of water that came down the Crocodile River caused a lot of damage and obviously financial stuff.

For example, the Sterkfontein caves and nearby properties in the CoH-WHS were temporarily closed in 2022 and 2023 due to flooding (Matshusa & Leonard, 2023). Excessive rainfall results in roads, bridges, and trails collapsing, and makes it difficult and unsafe to drive along the geoheritage routes in the geotourism destinations. These further increases the costs of rebuilding infrastructure before geotourism activities can resume. Participants also noted that encroachment of trees and invasive plants could damage the natural and aesthetic values of the geoheritage sites. Informant W, an official of one of the private reserves in the CoH-WHS and MBR, noted: "Encroachment of trees across the land. The western part of our reserve was quite impacted by the floods and our roads were washed away."

Damage to infrastructure such as roads, bridges, trails, and accommodation facilities can reduce the potential for geotourism because of the limited movement of tourists to geoheritage sites. Ground subsidence and groundwater flooding at surface level and below ground have been identified as serious hazards to geoheritage sites in protected areas such as the WHS of the UK (Cigna et al., 2018). Similarly, gradual degradation, changes in the frequency and severity of extreme events such as flooding, and seasonal changes present serious threats to the sustainability of geoheritage sites and geotourism (Gordon et al., 2022). These impacts do not only affect the scientific integrity and value of geoheritage sites but also lead to loss of income and employment (especially for poor communities who depend on the establishment for their livelihoods). The participants indicated that there were no legislation and policies in place to protect geoheritage sites. This is supported by Matshusa & Leonard (2023), whose research recommended the development of legal and governance instruments for the protection, preservation and management of geoheritage sites against nature-induced geoheritage threats (e.g. climate change) in protected areas such as national parks and WHS. Therefore, integrated management plans must include adaptation and resilience strategies to prevent or minimise the impacts of climate change on geoheritage sites and geotourism, including underground fossils that can be damaged by flooding and underground caving.

Climate impacts on geoheritage sites at the KNP

Some government participants indicated that they did not see any direct threat from climate change on geoheritage sites. Furthermore, community leaders and local residents at the KNP noted that they did not have any knowledge of the impacts of climate change on geoheritage sites and geotourism. These suggest a lack of capacity building and lack of knowledge on the impacts of climate change on geoheritage sites at the KNP. As noted by informant KM, a SANPark official, referring to the lack of knowledge on climate change:

> Now you have this white elephant called KNP and you have 6 million people living outside the park who have no idea what a national park is. How do we ensure that people come to love their national park in a way that will guarantee the long-term sustainability of that protected area? Educated people might say we need protected areas because it is good for biodiversity and climate change but if I am not [educated], it means nothing to me. . . . So, geotourism can help to transfer knowledge [on climate change].

This corroborates previous studies which found that geotourism can help to raise awareness of both the impacts of climate change on society and the environment as well as the impacts of society on the climate (Gordon, 2023). Field observation and interview findings show that geoheritage sites at the KNP were being damaged by erosion (Figure 4.4). Furthermore, field observations at the KNP indicated that geoheritage sites such as the Bobomeni waterfalls and some infrastructure (a beacon indicating the meeting point of the three countries, South Africa, Zimbabwe, and Mozambique) were destroyed by floods in 2000. Other observed impacts of extreme climate at the KNP included dinosaur bones with missing pieces near

Figure 4.4 The eroding sandstone boulders of the Mashikiri geoheritage site.

the Makahane geoheritage site and destroyed artefacts at the Makahane, Krem-etart, and Thulamela geoheritage sites (Matshusa et al., 2021a, 2021b) which were apparently washed away by water during strong wind and extreme rainy seasons. Furthermore, the rock art inside the cave and sandstone arc at the Mashikiri geoheritage site are exposed to damaging climatic conditions of high temperatures and floods. The KNP has been closed a few times because of flooding and fires. In this context, extreme climatic conditions such as floods, weathering, and high temperatures can damage geoheritage sites and tourism activities (Christian, 2018; Chikodzi et al., 2022; Matshusa & Leonard, 2023). As a result of severe climatic conditions, tourists may lose interest in visiting the tourism sites in the KNP (Steffen et al., 2009).

By merging the results of the three study sites, this study showed that various climate change impacts were associated with geoheritage sites in Southern Africa, especially at the CoH-WHS, MBR, and KNP. Climate change impacts have the potential to negatively impact geoheritage resources such as water resources (dams, lakes, rivers, streams, surface, and underground water), topographic elements (mountains and valleys), landscapes, the biosphere, flora, and fauna. By understanding the climatic conditions of the geoheritage sites, managers of protected areas can determine the suitable period for visiting the CoH-WHS, MBR, and KNP. Adaptation and monitoring climate strategies on the impacts of climate change on geoheritage sites can increase geotourism activities (Durlevic et al., 2021).

Conclusion

The overall objective of this study was to understand the impacts of climate change on geoheritage sites and geotourism in Southern Africa through a qualitative approach. The discussion of this study echoes the perspectives of key stakeholders and support previous studies. This study revealed that the impacts of climate change on geoheritage and geotourism have environmental, social, and economic implications. Extreme climatic conditions and events caused environmental damage to geoheritage sites in the study areas through weathering, erosion, flooding, pollution, and tree and plant encroachment. These have the potential to damage the scientific value, aesthetic value, and integrity of geoheritage sites and geotourism.

Flooding has also resulted in damage and collapse of infrastructure for geoheritage and geotourism. This led to secondary impacts such as pollution and sewage spilling into the rivers and other water bodies. These can cause social challenges such as health complications for human beings and animals from drinking contaminated water. Climate change exacerbates damage to infrastructure such as roads, bridges, trails, and accommodation facilities. This leads to the reduction of geotourism activities because of the limited movement of tourists to geoheritage sites. The study also revealed that climate change events such as floods and fires caused a lack of economic growth and loss of revenue and employment, especially for poor communities who depend on the establishments for their livelihoods. Hence, it can be concluded that the impacts of extreme climate change events have a negative impact on geoheritage sites and geotourism in Southern Africa, thereby affecting the environmental, social, and economic prospects of the region. Despite the limited growth of geoheritage and geotourism research in Southern Africa, climate change impacts are accelerating at a rapid rate.

In this context, legislation, policies, strategies and environmental plans on climate change adaptation must support tourism in view of extreme climate events such as droughts, fires and floods. This legislation and strategies on climate change must take the geoheritage sites in Southern African into account. Therefore, research on climate change impacts should be undertaken immediately to avoid the catastrophic destruction of geoheritage and geotourism in Africa. Capacity-building and awareness initiatives must be intensified to conscientise all the key stakeholders, especially in rural areas, about the impacts of climate change on geoheritage and geotourism. Such initiatives may include the development of undergraduate and postgraduate qualifications, workshops and short courses on geoheritage and geotourism. Understanding and mitigating the impacts of climate change on geoheritage sites and geotourism play a vital role in the protection and conservation of fauna, flora and their habitats.

By understanding the negative impacts of climate change on geoheritage sites in Southern Africa, this research contributes to reducing the harm to geoheritage sites and geotourism. The qualitative research revealed crucial knowledge and information that can assist in locating the potential impacts of climate change that may have become obscured. This study contends that it is essential to understand the perceptions of stakeholders to develop management and adaptation strategies to address the impacts of climate change on geoheritage and geotourism. This is vital for the selection of best management options to tackle threats to geoheritage sites and geotourism. Therefore, this study recommends that:

- A similar study should be undertaken for the African continent.
- Management and adaptation strategies are developed to provide guidelines for addressing the impacts of climate change on geoheritage and geotourism.
- Capacity-building and awareness initiatives are intensified to ensure that local communities and all key stakeholders have the necessary knowledge and skills to manage the impacts of climate change on geoheritage and geotourism.

References

African Development Bank. (2023). *Southern Africa economic outlook 2023*. www.afdb.org/en/documents/southern-africa-economic-outlook-2023 (accessed on 01 August 2023).

Aligishiev, Z., Massetti, E., Bellon, M. (2022). *Macro-fiscal implications of adaptation to climate change*. Washington, DC, United States of America: International Monetary Fund.

Bellon, M., Massetti, E. (2022). *Planning and mainstreaming adaptation to climate change in fiscal policy*. Washington, DC, United States of America: International Monetary Fund.

Carmona, J., Gil, T. (2022). UNESCO and the European Union come to the rescue of Doñana. In: S. Doempke, N. Bolomey, R.M.E.Saba, M.F. Oliveira, J.P.O Costa & S. Wagh (Eds.), *World heritage watch: World heritage watch report*, pp. 31–33. Berlin, Germany: World Heritage Watch.

Chalovska, E.A., Scarry, D., Dimoska, S. (2022). Management of the Lake Ohrid region: One step forward, two steps back. In: S. Doempke, N. Bolomey, R.M.E. Saba, M.F. Oliveira, J.P.O.Costa & S. Wagh (Eds.), *World heritage watch: World heritage watch report*, pp. 143–146. Berlin, Germany: World Heritage Watch.

Chikodzi, D., Nhamo, G., Dube, K., Chapungu, L. (2022). Climate change risk assessment of heritage tourism sites within South African National Parks. *International Journal of Geoheritage and Parks*, 10, 417–434. https://doi.org/10.1016/j.ijgeop.2022.08.007.

Christian, C.S. (2018). The Caribbean's geotourism potential and challenges: A focus on two islands in the region. *Geosciences*, 8, 1–12. https://doi.org/10.3390/geosciences8080273.

Cigna, F., Tapete, D., Lee, K. (2018). Geological hazards in the UNESCO world heritage sites of the UK: From the global to the local scale perspective. *Earth-Science Reviews*, 176, 166–194. https://doi.org/10.1016/j.earscirev.2017.09.016.

Creswell, J.W. (2014). *Research design: Qualitative, quantitative and mixed methods approaches*. Thousand Oaks, United States of America: Sage.

Crofts, R. (2019). Linking geoconservation with biodiversity conservation in protected areas. *International Journal of Geoheritage and Parks*, 7, 211–217. https://doi.org/10.1016/j.ijgeop.2019.12.002.

Crofts, R., Gordon, J.E., Brilha, J., Gray, M., Gunn, J., Larwood, J., Santucci, V.L., Tormey, D., Worboys, G.L. (2020). Guidelines for geoconservation in protected and conserved areas. In: C. Groves (Ed.), *Best practice protected area guidelines series no. 31*. Gland, Switzerland: International Union for the Conservation of Nature (IUCN).

Crofts, R., Tormey, D., Gordon, J.E. (2021). Introducing new guidelines on geoheritage conservation in protected and conserved areas. *Geoheritage*, 13, 1–14. https://doi.org/10.1007/s12371-021-00552-0.

Dube, K., Nhamo, G. (2018). Climate variability, change and potential impacts on tourism: Evidence from the Zambian side of the Victoria Falls. *Environmental Science and Policy*, 84, 113–123. https://doi.org/10.1016/j.envsci.2018.03.009.

Dube, K., Nhamo, G. (2020). Evidence and impact of climate change on South African National Parks: Potential implications for tourism in the Kruger National Park. *Environmental Development*, 33. https://doi.org/10.1016/j.envdev.2019.100485.

Durand, J.F., Meeuvis, J., Fourie, M. (2010). The threat of mine effluent to the UNESCO status of the cradle of humankind world heritage site. *The Journal for Transdisciplinary Research in Southern Africa*, 6(1), 73–92.

Durlevic, U., Cegar, N., Dobric, M., Vukašinovic, S., Lukic, T., Stevanovic, V., Radovanovic, D., Valjarevic, A. (2021). The heritage climate index (HERCI): Development, assessment and application for tourism purposes in geoheritage and cultural heritage sites. *Atmosphere*, 14, 1–18. https://doi.org/10.3390/atmos14081265.

Ehsan, S., Leman, M.S., Begum, R.A. (2013). Geotourism: A tool for sustainable development of geoheritage. *Advanced Materials Research*, 622, 1711–1715. https://doi.org/10.4028/www.scientific.net/AMR.622-623.1711.

El Hassi, M.F., Muftah, A.M. (2022). An inventory study of the geosites in the area from Susa to Darnah in Al Jabal al Akhdar, Libya: A proposed geopark. *Geoheritage*, 14, 122. https://doi.org/10.1007/s12371-022-00755-z.

Engelbrecht, F.A., Steinkopf, J., Padavatan, J., Midgley, G.F. (2024). Projections of future climate change in Southern Africa and the potential for regional tipping points. In: G.P. von Maltitz, G.F. Midgley, J. Veitch, C. Brümmer, R.P. Rötter, F.A. Viehberg & M. Veste (Eds.), *Sustainability of Southern African ecosystems under global change. Ecological studies 248*, pp. 169–190. Cham, Switzerland: Springer.

Gordon, J.E. (2023). Climate change and geotourism: Impacts, challenges and opportunities. *Tourism and Hospitality*, 4, 514–538. https://doi.org/10.3390/tourhosp4040032.

Gordon, J.E., Wignall, R.M.L., Brazier, V., Crofts, R., Tormey, D. (2022). Planning for climate change impacts on geoheritage interests in protected and conserved areas. *Geoheritage*, 14, 1–25. https://doi.org/10.1007/s12371-022-00753-1.

Hill, W., Kimball, S. (2023). Geoconservation and climate change working group established. *Geoheritage Newsletter*, 7, 3–4.

International Monetary Fund (IMF). (2022). *Regional economic outlook. Sub-Saharan Africa: Living on the edge*. Washington, DC, United States of America: IMF.

Kubalíková, L., Irapta, P.N., Pál, M., Zwoliński, Z., Coratza, P., Van Wyk de Vries, B. (2023). Visages of geodiversity and geoheritage: A multidisciplinary approach to valuing, conserving and managing abiotic nature. In: L. Kubalíková, P. Coratza, M. Pál, Z. Zwoliński, P.N. Irapta & B. van Wyk de Vries (Eds.), *Visages of geodiversity and geoheritage. Special publications 530*, pp. 1–12. London, England: Geological Society.

Magaliesberg Biosphere Reserve. (2021). www.magaliesbergbiosphere.org.za (accessed on 20 September 2021).

Matshusa, K., Leonard, L. (2023). Geoheritage threats in South African National Parks. *Tourism and Hospitality*, 4, 202–213. https://doi.org/10.3390/tourhosp4010012.

Matshusa, K., Thomas, P., Leonard, L. (2021a). A methodology for examining geotourism potential at the Kruger National Park, South Africa. *GeoJournal Tourism Geosites*, 34, 209–217. https://doi.org/10.30892/gtg.34128-639.

Matshusa, K., Leonard, L., Thomas, P. (2021b). The contribution of geotourism to social sustainability: Missed opportunity? *The International Journal of Social Sustainability in Economic, Social, and Cultural Context*, 17, 95–118. https://doi.org/10.18848/2325-1115/CGP/v17i01/95-118.

Mc Keever, P., Narbonne, G. (2021). *Geological world heritage: A revised global framework for the application of criterion (viii) of the world heritage convention*. Gland, Switzerland: IUCN.

Mehdioui, S., Hadi, H.E., Tahiri, A., Haibi, H.E., Tahiri, M., Zoraa, N., Hamoud, A. (2022). The geoheritage of northwestern Central Morocco area: Inventory and quantitative assessment of geosites for geoconservation, geotourism, geopark purpose and the support of sustainable development. *Geoheritage*, 14(86). https://doi.org/10.1007/s12371-022-00712-w.

Morriss, A. (2022). Will a waste incinerator threaten the Jurassic Coast? In: S. Doempke, N. Bolomey, R.M.E. Saba, M.F. Oliveira, J.P.O Costa & S. Wagh (Eds.), *World heritage watch: World heritage watch report 2022*, pp. 34–37. Berlin: World Heritage Watch.

Newsome, D., Dowling, R. (2018). Geoheritage and geotourism. In: E. Reynard & J. Brilha (Eds.), *Geoheritage: Assessment, protection and management*, pp. 305–321. Amsterdam, Netherlands: Elsevier.

Osipova, E., Emslie-Smith, M., Osti, M., Murai, M., Aberg, U., Shadie, P. (2020). *IUCN world heritage outlook 3: A conservation assessment of all natural world heritage sites*. Gland, Switzerland: IUCN.

Prosper, C.D., Diaz-Martinez, E., Larwood, J.G. (2018). The conservation of geosites: Principles and practice. In: E. Reynard & J. Brilha (Eds.), *Geoheritage: Assessment, protection and management*, pp. 417–430. Amsterdam, Netherlands: Elsevier.

Quesada-Valverde, M.E., Quesada-Román, A. (2023). Worldwide trends in methods and resources promoting geoconservation, geotourism, and geoheritage. *Geosciences*, 13(12), 1–12. https://doi.org/10.3390/geosciences13020039.

Reimold, W.U. (1999). Geoconservation – a Southern African and African perspective. *Journal of African Earth Science*, 29(3), 469–483. https://doi.org/10.1016/S0899-5362(99)00110-4.

Reynard, E., Brilha, J. (2018). Geoheritage: A multidisciplinary and applied research topic. In: E. Reynard & J. Brilha (Eds.), *Geoheritage: Assessment, protection, and management*, pp. 3–9. Amsterdam, Netherlands: Elsevier.

Rogerson, C.M., Van der Merwe, C.D. (2016). Heritage tourism in the global south: Development impacts of the cradle of humankind world heritage site, South Africa. *Local Economy*, 31(1–2), 234–248.

Schluter, T., Schumann, A. (2018). Geosites as a potential for the development of tourism – overview of relevant sites in Eswatini (formerly Swaziland). *GeoJournal of Tourism and Geosites*, 22(2), 535–547. https://doi.org/10.30892/gtg.22222-309.

Selmi, L., Canesin, T.S., Gauci, R., Pereira, P., Coratza, P. (2022). Degradation risk assessment: Understanding the impacts of climate change on geoheritage. *Sustainability*, 14(7), 1–19. https://doi.org/10.3390/su14074262.

Shapi, M., Jordaan, M.A., Mbambo, A.T., Davies, T.C., Chirenje, E., Dube, M. (2021). Determination of potentially harmful element (PHE) distribution in water bodies in Krugersdorp, a mining city in the West Rand, Gauteng province, South Africa. *Minerals*, 11, 1–19. https://doi.org/10.3390/min11101133.

Sharples, C. (2002). *Concepts and principles of geoconservation*. Hobart: Tasmanian Parks and Wildlife Service. www.dpiw.tas.gov.au/inter.nsf/Attachments/SJON-57W3YM/$FILE/geoconservation.pdf (accessed on 11 June 2018).

Steffen, D., Schlunegger, F., Preusser, F. (2009). Drainage basin response to climate change in the Pisco Valley, Peru. *Geology*, 37(6), 491–494. https://doi.org/10.1130/G25475A.1.

Štrba, L., Vravcová, A., Podoláková, M., Varcholová, L., Kršák, B. (2023). Linking geoheritage or geosite assessment results with geotourism potential and development: A literature review. *Sustainability*, 15(12), 1–20. https://doi.org/10.3390/su15129539.

United Nations Educational, Scientific and Cultural Organization (UNESCO). (2018). *Global geoparks*. www.unesco.org/new/en/natural-sciences/environment/earth-sciences/unesco-global-geoparks/ (accessed on 11 June 2018).

United Nations Educational, Scientific and Cultural Organization (UNESCO). (2023). *Climate change vulnerability and risk in UNESCO global geoparks and biosphere reserves in Latin America and the Caribbean: Executive summary.* https://unesdoc.unesco.org/ark:/48223/pf0000388007 (accessed on 31 January 2024).

Woo, K.C., Worboys, G. (2019). Geological monitoring in protected areas. *International Journal of Geoheritage and Parks*, 7, 218–225. https://doi.org/10.1016/j.ijgeop.2019.12.004.

Zhang, J., Xiong, K., Liu, Z., He, L. (2022). Research progress on world natural heritage conservation: Its buffer zones and the implications. *Heritage Science*, 10(102), 1–21. https://doi.org/10.1186/s40494-022-00744-z.

5 Unveiling Southern Africa's geotourism potential

Analysing the political economy of mining development and its effects on geotourism

Llewellyn Leonard

Introduction

The United Nations Educational, Scientific and Cultural Organization (UNESCO) World Heritage Centre has identified mining development as a significant threat to the conservation and management of various World Heritage properties globally. In 2015, out of 141 World Heritage properties that submitted conservation reports, 18 per cent reported being affected by mining activities. Distressingly, among these affected properties, 84 per cent were cultural landscapes or natural sites or a combination of both (Malijani, 2020). Southern Africa is a treasure trove of geological and heritage wonders, boasting a rich tapestry of landscapes (Norman, 2013) and a botanical gem of roaming sceneries (Cleveland, 2021). These wonders have captivated the imaginations of travellers and geologists alike. From the rugged cliffs of the Drakensberg Mountain Range, which encloses the central Southern African plateau; the Oribi Gorge, a canyon in southern KwaZulu-Natal; the Victoria Falls, bordering Zambia and Zimbabwe; to the ancient rock formations of the Namibian Desert, Southern Africa is endowed with geological features that serve as a testament to the Earth's remarkable history and natural heritage. Unfortunately, in addition to these natural wonders, globally and at the regional and local scales in many countries, mineral mining may result in significant impacts on natural resources such as water due to high-energy industries' dependence on this resource (Meißner, 2021). Water plays a pivotal role in shaping geological landscapes and contributes significantly to the appeal and sustainability of geotourism destinations. According to Worlanyo & Jiangfeng (2021), mining operations can continue to affect ecosystems long after operations cease, with loss of vegetation cover, mass destruction of water sources, loss of biodiversity, land-use changes and food insecurity, increased social vices and conflicts, high cost of living, and air pollution to name a few. Hence, there is a cycle of mining development and impacts, can in turn affects the development of geotourism.

Before proceeding, it is important to clarify what is meant by the term "geotourism". Although there are many understandings of geotourism, it was initially understood narrowly as a form of sustainable tourism that centres on the appreciation and conservation of geological features (Dowling, 2013; Ruban, 2015).

DOI: 10.4324/9781032706764-5

A more comprehensive, Eurocentric understanding of geotourism may be taken from the European Geoparks Network (2011), which states that it is a form of tourism that sustains and enhances the identity of a territory, taking into consideration its geology, environment, culture, aesthetic, and heritage, as well as the well-being of residents. This understanding is more comprehensive in that it includes features of a territory beyond just geomorphology. However, geotourism is a relatively new concept that has been underdeveloped (Anhaeusser et al., 2016), with limited literature on the subject (especially for sub-Saharan Africa) (Ólafsdóttir & Tverijonaite, 2018). However, in Africa, and specifically Southern Africa, literature is scarce, albeit with recent contributions from South Africa (e.g. Matshusa & Leonard 2023; Matshusa et al., 2021; Taru & Chingombe, 2016; Mukwada & Sekhele, 2017). In the context of Southern Africa, geotourism holds profound significance and can serve as a magnet for tourists seeking to unravel and explore the uniqueness of a region and its heritage. Moreover, geotourism contributes significantly to local economies, supporting livelihoods and cultural preservation, as well as fostering community engagement (Newsome et al., 2012).

However, Southern Africa's potential as a geotourism hub faces multifaceted challenges, particularly with the pervasive influence of the mining industry. Mining activities, which have been deeply embedded in Southern Africa's economic fabric for decades, have left indelible imprints on the environment, cultural heritage, and local communities (Magadza & Zinyama, 2007). There is a notable gap in studies examining the repercussions of mining on (geo)tourism and conservation sites in South Africa (Leonard, 2016). A Scopus database search in April 2024 to explore journal article titles with the keywords "mining", "geotourism", and "Africa" produced no results, although one conference paper was found. However, article titles with the keywords "mining", "tourism", and "Africa" produced four results, showing that the link between mining and geotourism is an underexplored area. Thus, in this study a search was done for popular media and other open-source documents to investigate mining impacts on geotourism development. However, journal papers dealing with geotourism in Africa were used as required. A further Scopus database search to explore journal articles with the keywords "geotourism" and "Africa" produced 25 results (Table 5.1). This search revealed that most geotourism literature emanated from South Africa (11 papers), followed by North Africa (5 papers). A large number of journal titles were found containing the keywords "mining" and "Africa" (780), showing that Africa's natural resources are under siege from mining activities. A summary of the database findings is presented in Table 5.1.

The exploration and exploitation of mineral resources for economic growth can clash with the principles of geotourism, posing threats to geological sites and their associated environmental and socio-cultural contexts. The collision between the mining sector and the burgeoning potential of geotourism presents a complex dynamic that shapes the region's landscapes, cultural heritage, and prospects for sustainable tourism development. Mining ecological footprint, alteration of landscapes, and resource depletion threaten the integrity of geological formations,

Table 5.1 Title and journal article database search linking mining and (geo)tourism in Africa.

Database search for journal article titles with the keywords "mining" and "geotourism" and "Africa"	*Article titles with the keywords "mining", and "tourism" and "Africa"*	*Articles with the keywords "geotourism" and "Africa*	*Articles with the keywords "mining" and "Africa"*
0 results	4 results (All titles from South Africa)	25 results (Africa 3 papers; Southern Africa 12 papers (South Africa predominantly); West Africa 2 papers; North Africa 5 papers; East Africa 1 paper; Central Africa 2 papers)	760 results

deterring their potential as geotourism attractions. Furthermore, mining's socioeconomic impacts on local communities, cultural heritage, and traditional knowledge systems cast shadows on the region's geotourism potential. As noted by Malijani (2020), during mining development, substantial archaeological and palaeontological sites and human remains often surface due to the extensive history of human habitation across various African cultural landscapes. Unfortunately, the land transformation involved in mining activities poses risks to these invaluable cultural treasures. Such transformations often result in damage to cemeteries, sacred sites, and other significant heritage features, causing adverse effects on local communities and their cultural heritage.

Examining the intersection between mining development and geotourism in Southern Africa becomes imperative. This chapter aims to unravel the complexities surrounding this intricate relationship, exploring the impacts of mining on geological landscapes, cultural heritage and local economies, and endeavouring to chart a path towards sustainable geotourism development amid the dominant influence of the mining industry. By investigating governance frameworks and sustainable development approaches, it seeks to unveil potential pathways for sustainable geotourism regional development. It advocates for inclusive decision-making processes and equitable distribution of benefits to foster ecological justice and geotourism development. Within this context, this chapter is divided into several sections. The next section explores the interplay between mining, biodiversity, and sustainable development in Southern Africa, followed by mining impacts on archaeology and cultural heritage because these impacts are largely irreversible. The chapter then explores the governance of mining in Southern Africa and its implications for geotourism. The final section contains discussions and conclusions.

Mining, biodiversity and sustainable development in Southern Africa

Comprehensive studies assessing the impacts of mining on biodiversity and sustainable development in Southern Africa, and indeed across the African continent, have been lacking notably. Even though some research on the impact of mining on (eco)tourism exists globally, it remains limited. While limited global research has explored the correlation between mining impacts on natural environments (see e.g. Dwyer et al., 2014; Folkersen et al., 2018), in South Africa, limited studies explored mining impacts on tourism and natural sites in places such as Dullstroom, Mpumalanga (Leonard, 2016), acid mining drainage (AMD) and impacts on nature conservation in the West Rand, Johannesburg (Leonard & Langton, 2016), and mining development impacts in the Mapungubwe World Heritage Site tourist attraction in Limpopo (Leonard & Lebogang, 2018), to name a few. However, investigations focusing specifically on the interplay between industrial activities like mining and impacts on geotourism more widely in Southern Africa are lacking. Some recent notable work on geotourism generally in Southern Africa was conducted by a few authors exploring geotourism in South Africa (Matshusa & Leonard, 2023); geotourism and archaeology in Tunisia (Matoussi et al., 2023); geotourism and landscapes in East Africa (Scoon, 2020); geological heritage in Egypt (Sallam et al., 2020); and rock art conservation and geotourism in Morrocco (Abioui et al., 2019). The majority of these studies originate from South Africa. There is an urgent need for more research on the impacts of mining on geotourism sites, given that mining has and will likely continue to negatively affect natural environments with significant geotourism potential as well as sustainable development.

Mining growth in Southern Africa and socio-environmental impacts

Rich in mineral wealth, countries such as Zimbabwe, Ghana, the Congo, and Ethiopia have and will continue to harness their geological endowments, particularly gold, diamonds, platinum, and other precious minerals driving economic growth and attracting foreign investment (Cust & Zeufack, 2023; Ngang & Kamga, 2022). Mining development comes at a cost due to conflict and the destruction of environmental resources. For example, according to Kumah & Nyarko (2018), several resource conflict scenarios have arisen across Africa, primarily due to the control and exploitation of valuable resources such as diamonds, oil, and gold. These instances include the notorious “blood diamond” incidents witnessed in Sierra Leone, Angola, and Côte d’Ivoire. Additionally, the Niger Delta has been a focal point of conflicts stemming from oil exploitation in southern Nigeria. Other emerging conflicts involve gold exploitation, notably the *galamse* (unlicensed artisanal gold miners) in Ghana and the *orpailleurs* (artisanal gold miners in Francophone West African countries like Mali and Burkina Faso). However, this has been to the detriment of the African continent, its natural environment, and its people (Denhere, 2021), including geotourism development. The mining industry’s operations often intersect with areas of high conservation and geology value, creating tension between economic gain and the preservation of the natural environment (Agaba, 2022). Mining activities often result in substantial land degradation, such as the formation of open pits and extensive rock waste dumps, leading to severe environmental repercussions that can endanger plant and animal species reliant on specific environmental conditions for survival (Muma et al., 2020). Of concern is that post-mining impacts can persist, with heavy metals and corrosive substances leaching into waterways and the environment, impeding the recovery of wildlife and vegetation in the region (Cacciuttolo et al., 2023).

Mining activities, characterised by excavation, land disturbance, and alterations to landscapes, can directly impact (geo)tourism sites, potentially diminishing their appeal to tourists (Newsome et al., 2012; Ngang & Kamga, 2022). For example, the geoheritage of the Central Spanish Volcanic Region (Ciudad Real) has been significantly affected by open-pit mining since the beginning of the twentieth century, with some 30 volcanoes destroyed (Poblete Piedrabuena et al., 2019). Similarly, large-scale open-pit mining or quarrying activities have transformed natural landscapes in Southern Africa, altering the aesthetic and cultural significance and sense of place. In South Africa, AMD caused by diamond and gold mining during apartheid left a toxic trail of historical pollution that has continued into the new democracy, impacting the natural environment (Laker, 2023). Mining developments are being approved by the ruling party via the national Department of Mineral Resources (DMR) in sensitive and protected environmental areas (Leonard, 2016). Moreover, mining licenses have been granted or are pending in sensitive areas such as the Mapungubwe World Heritage Site, Blyde River Canyon near the Kruger National Park, the KwaZulu-Natal Coast, and the Wild Coast in the Eastern Cape (Leonard & Lebogang, 2018). Additionally, mining licenses can have impacts in protected sites used by indigenous communities and livelihoods. A case illustrating this issue occurred on 28 May 2015 in South Africa when the Traditional Healers Organisation pressed criminal charges against Aquila Steel SA (Pty) Ltd and three of its directors. They were accused of unlawfully causing extensive damage to a highly revered cultural site in South Africa. Aquila Steel undertook unauthorised activities such as clearing over 33 kilometres of roads across the Madimatle Mountain and drilling in roughly 200 locations, leading to the illegal removal of vegetation and protected tree species. This case highlights a larger issue of the DMR granting prospecting and mining rights inappropriately and without proper rationale (Davis, 2015). Such actions pose threats to the country's geotourism potential.

Nyamekye & Mensah (2012) note how mining operations over many years have impacted the environment and communities in Ghana. For example, in 2001 Goldfields Limited spilled cyanide into the Asuman River, killing birds, crabs, and fish. This river provides drinking water to communities in Abekoase et al. Therefore, mining can significantly impact and change the natural environment and have implications for biodiversity, including health and well-being. There is no doubt that mining will have far-reaching impacts on Southern Africa's natural environments. A notable example is the Republic of Guinea, where the Mount Nimba Biosphere Reserve (a designated World Heritage Site) was reduced by 1,550 hectares to accommodate iron-ore prospecting. Unfortunately, a substantial 44 per cent of Africa's major metal mines are situated within a 10-km radius of protected areas, which is significantly higher than the 25 per cent observed in both Asia and South America (Edwards et al., 2014). Thus, mining poses significant threats to the long-term sustainability of parks and their geotourism assets, as mining (being short-term and environmentally destructive) contrasts sharply with the longevity and sustainability of geotourism.

Mining impacts on archaeology and cultural heritage

Mining operations often disrupt indigenous communities and their connections to geological landscapes and archaeological heritage. Indigenous knowledge systems linked to geological features (including stories, rituals, and traditional practices associated with the environment) face erosion due to mining-related alterations (Magadza & Zinyama, 2007). Notably, African World Heritage cultural landscapes face the highest risk and threats from ongoing and proposed mining and exploration ventures (Malijani, 2020). Mining-induced displacement and socioeconomic changes in local communities may disrupt traditional livelihoods and heritage-based tourism initiatives, undermining the region's archaeology and geotourism. In Ghana, Kumah & Nyarko (2018) conducted a study on the impact of mining on cultural heritage in the Prestea area, notably Awudua Dada. The archaeological survey of ancient gold mines revealed the discovery of cultural heritage items such as pots, stone tools, and glass bottles used by earlier settlers in the region. The artefacts were retrieved from ancient mine pits nearby, evidencing their proximity. Unfortunately, due to digging by miners, some of these materials were partially or entirely destroyed. In their study, Leonard & Lebogang (2018) delved into the repercussions of mining on local sustainability, focusing on the Mapungubwe Heritage Site in Limpopo. The findings of the study revealed that mining companies and the government exhibited a preference for inadequate public participation and engaged in corrupt practices over the interests of the mining sector. Notably, the Mapungubwe Heritage Site is among South Africa's seven World Heritage Sites and encompasses the archaeological marvel of Mapungubwe, a kingdom predating the renowned Great Zimbabwe.

Similarly, historical pollution and pollution from new mining developments in South Africa have led to the contamination of archaeological sites. This has already been witnessed in the Randfontein area, where AMD flows northwards towards the Cradle of Humankind World Heritage Site (CoHWHS) (Ochieng et al., 2010), which is a UNESCO-protected site. Over a decade ago, Durand et al. (2010) already warned of significant environmental risk posed by AMD to archaeological and palaeontological resources at the CoHWHS, situated adjacent to one of the richest gold-bearing geological sequences in the world. These authors noted that AMD threatened the biggest collection of hominin fossils and had a major impact on the archaeological and palaeontological heritage of not only South Africa but the world, including the tourism, hospitality, and education sectors of South Africa. Based on their research, Hobbs & Mills (2012) also reported the CoHWHS being under threat from AMD due to groundwater contamination putting fossil and archaeological sites at risk.

Regarding an evaluation of heritage impact conducted by eThembeni Cultural Heritage (2007), the proposed Xolobeni Mineral Sands Project along the Wild Coast in the Eastern Cape province of South Africa was eventually halted. This was after a legal victory and court case brought by the community against the DMR and the mining company as the direct repercussions of the proposed mining on the cultural and archaeological significance of the region would be devastating because the tribal area has substantial archaeological value; and graves and ancestral burial sites were at risk of being disturbed or destroyed (Ellis, 2020). The proposed mining areas were to encompass the immediate coastal littoral of Plio-Pleistocene palaeo-dunes overlain with more recent Late Pleistocene-Holocene aeolian sands. These contain the fossil rich Mzamba Cretaceous Formation. As such, mining can result in irreversible harm to archaeological sites and cultural heritage, and obliterating evidence related to past human history (eThembeni Cultural Heritage, 2007).

However, studies have also shown that in addition to (eco)tourism and other recreational activities offered by destinations, abandoned mines and the diverse geology, geomorphology, and archaeology exposure due to mining activities can play a role in (geo)tourism potential where post-mining sites become historical tourism attractions (see, e.g., Conlin & Jolliffe, 2011; Coupland & Coupland, 2014; Alonso, 2016). For example, Mir et al. (2023) researched the geoheritage and geotourism potential of the Kalaruch ancient mines and Satbarren archaeological stone in the Himalayas, emphasising their significance for preservation, conservation, and sustainable development. The investigation revealed that these unique structures hold immense geological, archaeological, historical, cultural, educational, and scientific importance. The surrounding area, characterised by beautiful topography and geomorphic landforms adorned with lush green forests, presents marvellous potential for geotourism development. Consequently, transforming this site into a geoheritage and geotourism destination could contribute to enhancing the socioeconomic status of the local population and ensure its preservation for future geologists, archaeologists, researchers, students, and curious tourists.

Similarly, Singh & Ghosh (2019) emphasize that geotourism at mining sites is becoming popular in some countries, as this results in the preservation of mining heritage and brings extra revenue. The authors note for an open-cast mine in eastern India that many attributes can be projected as geotourism assets. By designing human-made landscapes to mimic mountains and valleys, showcasing various stages of plant succession on dumping grounds, displaying fossils, and demonstrating mining activities using different machinery types, the site can be promoted and attract visitors. This approach can pave the way for successful mining geotourism development. Nevertheless, considering the largely negative impacts of mining on the environment and society, including the destruction of archaeology and cultural heritage (as highlighted earlier), mining development is unsustainable. However, already abandoned mining can be converted into geotourism sites such as in Iran (see Ghazi et al., 2021) where mining sites have geo-scientific, industrial, cultural, social, economic, historical, and anthropological potential for the development of (geo)tourism. Iran's rich mining heritage provides a valuable basis for (geo)tourism development and creates potential for diverse employment opportunities.

Governance of mining in Southern Africa and implications for geo(tourism)

Mining operations and developments in Southern Africa continue to negatively affect people and the environment, largely due to poor governance and corruption that have implications for geotourism in the region. Extensive financial inflows from mining activities, coupled with poor national governance, contribute to ubiquitous corruption. Several African nations experiencing rapid growth in mining and related infrastructure projects exhibit low rankings on Transparency International's corruption perceptions index (Transparency International, 2024). Effective governance prioritizes the management of industrial risks, particularly in sectors like mining, and acknowledges their environmental repercussions (Leonard & Lidskog, 2021). A crucial aspect of good governance involves empowering citizens to express their opinions and facilitating their participation in decision-making processes (Orr, 2002). Stirling (2006) categorises governance into "unreflective governance" that is characterised by limited decision making driven by instrumental motives and "reflective governance" that involves critical efforts to manage side effects and incorporate diverse perspectives for the implementation of sustainable practices. Contemporary governance approaches commonly emphasize the involvement of non-state stakeholders in decision-making processes (Van Marissing, 2005). These approaches underscore the importance of accountability, transparency, fairness, adherence to the rule of law, and ethical considerations by the state (McNutt & Rayner, 2010). Importantly, they advocate for a departure from relying solely on technocratic and bureaucratic processes in development management (Wesselink et al., 2011).

Around 30 per cent of the world's mineral resources are situated in Africa, making the continent a focal point for mining investment and development. This significant reserve has garnered attention from both developed and developing nations, including Australia, Brazil, Russia, India, and China (United Nations Environment Programme, 2021). The surge in African mining activities since the beginning of the century has drawn billions of United States dollars in investment (Janneh & Ping, 2011). However, this influx of investment has brought about environmental degradation and "unreflective governance", accompanied by the depletion of conservation efforts and hindrances to (geo)tourism development. This is particularly evident as foreign mining companies in Southern Africa engage in infrastructure development in exchange for access to natural resources (World Bank, 2019). The degradation of natural resources and hence the region's geotourism potential is, in part, associated with corrupt practices. Human Rights Watch (2017) observes that in general nations endowed with mineral resources, including those rich in oil, grapple with issues of corruption. In Ghana, as in many other African countries, the government may prioritise large-scale projects such as mining, with little or no attention given to how these policies might impact cultural heritage and the environment (Leonard, 2023). Knutsen et al. (2017) investigated how mining affected local corruption in Africa. The authors found that mining increased bribe payments. While mines are initially located in less corrupt areas, mining areas turn more corrupt after mines open and enable local officials to acquire bribes.

Due to large-scale corruption and lack of regulations and/or poor enforcement, some civil society groups have worked to hold mining industries and governments to account to protect natural environments for tourism and geotourism. As highlighted in the previous section on the impacts of mining on archaeology and cultural heritage surrounding the previously proposed Xolobeni Mineral Sands Project along the Wild Coast in the Eastern Cape and the cultural and archaeological significance of the region, like other sites this proposed site for mining development was accompanied by poor governance and corruption that resulted in an outcry from communities and civil society groups. For example, in April 2018, rural residents (consisting of 128 subsistence farmers and other inhabitants) of Pondoland initiated a legal test case in the Pretoria High Court seeking permission from mining companies to extract minerals. In November 2018, a court judgement emphasised the need for thorough consultation with the Pondoland community before granting any mining rights (Leonard, 2019). Despite some tribal leaders supporting mining development due to corruption and for its associated benefits (Pearce, 2017), the majority of Pondoland's residents argued that protecting the natural environment and tourism presented a more viable option than any mining development (Pretorius, 2016). The mining dispute caused fractures within the community, leading to violence and even fatalities (Seccombe, 2018). Additionally, concerns have been raised about corruption in the DMR and the mining company (Davis, 2013). Nevertheless, despite the evident impact of mining on the natural environment and the outcry from civil society, mining developments continue to receive government approval. Regrettably, mechanisms for democratic participation have not effectively enhanced participatory governance in this context (Leonard, 2017).

Such conflicts highlight the complex interplay between economic interests, environmental concerns and community well-being in the context of mining developments in nature tourist areas and implications for geotourism. Illustratively, the Zimbabwean government once awarded coal mining concessions in the Hwange National Park to multiple Chinese companies. This move resulted in substantial backlash from environmentalists, who expressed concerns that the mining activities could lead to environmental degradation and exacerbate human–wildlife conflict. Civil society advocated for the government to prioritize tourism and wildlife conservation in the park over mining endeavours (Karombo, 2020). These calls proved effective, prompting the Zimbabwean government to declare a ban on all mining activities in the park (BBC News, 2020). Similarly, the Save Zambezi coalition praised the Zambia Environmental Agency (ZEMA) for reversing its decision to permit copper mining in the Lower Zambezi National Park. ZEMA's decision effectively halted the proposed mining operations by Mwembeshi Resources Limited. Various civil society organisations, including the World Wildlife Fund, Conservation Advocates Zambia, and Conservation Lower Zambezi, diligently raised awareness about the detrimental effects that the copper mine would have on the park, the river, and the ecosystems and communities who were reliant on them. Specifically, the mine would have caused substantial pollution and negative impacts on the river, endangering its freshwater fisheries, which are crucial for the food security and livelihoods of communities in Zambia's Luangwa province and downstream areas in Zimbabwe and Mozambique (WWF, 2023). These cases illustrate the need for citizens and communities to hold governments accountable to ensure that mining development is not conducted in isolation; that a holistic assessment is done that considers the broader impacts on communities, society, the environmental, and cultural heritage; and that geotourism development is taken into account.

Discussion and conclusion

This chapter sheds light on the complex interplay between mining development and geotourism in Southern Africa. The undeniable economic significance of the mining industry poses challenges to the preservation of geological wonders that are essential for geotourism. This is considering that Southern Africa is rich in mineral wealth and its countries will continue to harness their geological endowments, particularly in gold, diamonds, platinum, and other precious minerals, driving economic growth and attracting foreign investment. The exploration of the impact of political economy industrial processes, specifically mining development, on geotourism sites and local economic development remains relatively uncharted territory in academic discourse. This chapter sought to address this gap, focusing on the Southern Africa region that is renowned for its rich geological features with the potential for geotourism development. The dominant force of the mining industry in the region has raised concerns about its potential adverse effects on conservation, archaeology, cultural heritage, the environment generally, and local communities. The findings reveal multifaceted impacts of mining on geotourism. Natural resources, vital for geotourism appeal, face depletion and alteration due to mining activities. Archaeological and cultural heritage, intricately linked with geological features, are at great risk from mining activities.

Additionally, local economies and communities, often intertwined with (geo)tourism, stand to experience losses in livelihoods and long-term employment opportunities due to mining destruction of sites with geotourism potential. The decline in job opportunities within the mining sector signals an urgent need to explore new sustainable markets, notably geotourism, for local economic development and employment generation. For example, Panchia (2023) notes that in South Africa, the plummeting prices of platinum-group metals, compounded by logistical and energy crises, are not just economic problems but are morphing into a national concern with far-reaching implications for employment and social stability. Therefore, the shift to enable geotourism in Southern Africa is crucial to also ensure a new tourism market offering that will contribute to local social sustainability, improving life on land and well-being. Sustainable communities and life on land are pivotal objectives of the global sustainable development agenda, and necessitate a strategic and responsible approach in geotourism-rich areas. Thus, the impact of mining on geotourism in the region is not solely an environmental concern; it is inherently linked to the achievement of broader sustainable development goals (e.g. zero hunger, good health and well-being, and clean water). Understanding the dynamics between any mining development and geotourism is crucial to devising strategies for sustainable geotourism and the protection of cultural heritage.

Regrettably, Southern Africa has grappled with the detrimental consequences of poor governance and widespread corruption, particularly in the context of mining development. This has implications for geotourism and its development in the region. Disturbingly, corruption has been identified across African countries, reflecting notably high rankings on the Transparency International index. This pervasive corruption extends to the mining sector, with literature indicating a concerning escalation in bribe payments to officials, particularly in regions associated with mining activities. A concerning trend has been observed, indicating that corruption tends to intensify in areas where mining operations are established, creating an environment where ethical concerns are often compromised after the commencement of mining activities. This interplay between corruption and mining poses a significant threat to the preservation of geotourism sites, further highlighting the urgent need for robust governance structures, anti-corruption measures, and ethical practices in the management of mining activities across the Southern African region. Unfortunately, an influx of mining investment has brought about environmental degradation and "unreflective governance", accompanied by the depletion of conservation efforts and hindrances to (geo)tourism development. Addressing governance challenges is essential for sustaining both the economic benefits derived from mining and the ecological integrity of the geotourism sites crucial for Southern Africa's unique appeal.

Across Southern Africa, governance structures need to be strengthened to ensure responsible mining practices and effective regulations that safeguard geotourism sites. Since mining is not set to disappear, balancing economic growth with environmental conservation and social well-being is imperative for the region's unique geological heritage and its socioeconomic future. This also requires that governments balance economic interests with the preservation of geological landscapes and cultural heritage for sustainable geotourism growth in the region. The responsible authorities must create platforms to involve citizens in decision-making structures. Such actions will result in governance shifting from "unreflective governance" to "reflective governance", which involves critical efforts to manage side effects and incorporate diverse perspectives for the implementation of sustainable practices. Contemporary governance approaches commonly emphasize the involvement of non-state stakeholders in decision making. Additionally, since corruption is a major concern, implementing and reinforcing transparent governance structures and anti-corruption measures specifically tailored for the mining sector are vital. This includes stringent monitoring and accountability mechanisms to curb unethical practices.

Addressing the above governance challenges, including mining and geotourism challenges, necessitates a nuanced approach that emphasises stakeholder collaboration with local communities and effective policies implemented to protect geotourism interests for the benefit of the region. The responsible authorities must prioritise community involvement in decision-making processes related to mining development. Communities must foster a sense of ownership of the land, since they live off the land and the land may offer immense geotourism potential. Considering that poor governance and corruption in mining development pervade the Southern African landscape, communities and civil society groups have an important role to play in holding governments and mining companies accountable. Despite the inherent government apathy to protect tourism sites and enforce regulations, civil society does have the potential to ensure the implementation of regulations, halt unjust mining development, and protect geotourism sites (as witnessed in Xolobeni, South Africa). However, to be more effective in ensuring accountability, civil society needs to work within and across class, race, and national lines to learn from each other and prioritise (geo)tourism to ensure employment and long-term sustainability.

This chapter underscored the imperative of a critical assessment of the nexus between mining development and geotourism to foster sustainable tourism and achieve overarching sustainable development goals. It advocates for conscientious governance and alternative development paradigms in the mining sector to safeguard the region's geological heritage and promote equitable and ecologically responsible geotourism. By juxtaposing the impacts of mining development against the aspirations of geotourism and sustainable development goals, this chapter highlights the urgency of recalibrating and formulating policies and practices. It emphasises the need for collaborative efforts among stakeholders, encompassing governments, industries, local communities and conservationists, to pave the way for a sustainable geotourism future in Southern Africa. For comprehensive and inclusive development across African nations, it is imperative to deeply consider the cultural aspects in major national decisions about geotourism. As Kumah & Nyarko (2018) note, even archaeologists can play a crucial role in this endeavour. This inclusive approach can provide valuable insights for resolving mining and geotourism conflicts.

References

Abioui, M., M'barki, L., Benssaou, M., Ezaidi, A., El Kamali, N. (2019). Rock art conservation and geotourism: A practical example from Foum Chenna engravings site, Morocco. *Geoconservation Research*, 2(1), 1–11. https://doi.org/10.30486/gcr.2019.664488.

Agaba, N. (2022). Rationale for Africa's legitimate ownership of her natural resources. In: Ngang, C. & Kamga, S. (Eds.), *Natural resource sovereignty and the right to development in Africa*. New York: Routledge.

Alonso, C. (2016). New tourist's uses for the mining heritage in Spain. *PASOS: Revista de Turismoy Patrimonio Cultural*, 14(4), 1013–1026.

Anhaeusser, C.R., Viljoen, M.J., Viljoen, R.P. (2016). *Africa's top geological sites*. Cape Town: Struik Nature.

BBC News. (2020). *Zimbabwe bans coal mining in Hwange and other game parks*, 9 September. www.bbc.com/news/world-africa-54085549 (accessed on 17 June 2021).

Cacciuttolo, C., Cano, D., Custodio, M. (2023). Socio-environmental risks linked with mine tailings chemical composition: Promoting responsible and safe mine tailings management

considering copper and gold mining experiences from Chile and Peru. *Toxics*, 11(5), 462. https://doi.org/10.3390/toxics11050462.
Cleveland, T. (2021). *Exoticism, exploitation and enrichment: A history of tourism in Africa*. Ohio: Ohio University Press.
Conlin, M., Jolliffe, L. (2011). *Mining heritage and tourism: A global synthesis*. New York: Routledge.
Coupland, B., Coupland, N. (2014). The authenticating discourses of mining heritage tourism in Cornwall and Wales. *Journal of Sociolinguistics*, 18(4), 495–517.
Cust, J., Zeufack, A. (Eds.). (2023). *Africa's resource future: Harnessing natural resources for economic transformation during the low-carbon transition*. Washington, DC: World Bank Publications.
Davis, T. (2013). Miners thumb noses at ecosystems. *Business Day*, 22 November.
Davis, T. (2015). Mining – coming to a protected area near you. *Ground Up*, 6 July.
Denhere, R. (2021). Evaluation of beneficiation constraints for identified critical minerals in Zimbabwe. Doctoral dissertation, Faculty of Engineering and the Built Environment, University of the Witwatersrand, Johannesburg.
Dowling, R.K. (2013). Geotourism: A global success story. *Geographical Review*, 103(3), 297–300.
Durand, J., Meeuwis, J., Fourie, M. (2010). The threat of mine effluent to the UNESCO status of the cradle of humankind world heritage site. *The Journal for Transdisciplinary Research in Southern Africa*, 6(1), a125. https://doi.org/10.4102/td.v6i1.125.
Dwyer, L., Pham, T., Jargo, L., Bailey, G., Marshall, J. (2014). Modelling the impact of Australia's mining boom on tourism. *Journal of Travel Research*, 55(2), 233–245.
Edwards, D.P., Sloan, S., Weng, L., Dirks, P., Sayer, J., Laurance, W.F. (2014). Mining and the African environment. *Conservation Letters*, 7, 302–311. https://doi.org/10.1111/conl.12076.
Ellis, E. (2020). Game-changing Xolobeni judgment orders applications for mining licences to be made public. *Daily Maverick*, 14 September.
eThembeni Cultural Heritage. (2007). *Heritage impact assessment of Xolobeni mineral sands project, Eastern Cape Province, South Africa*. https://sahris.sahra.org.za/sites/default/files/heritagereports/HIA_xolobeni_Van_Schalkwyk_LO_Oct07_0.pdf (accessed on June 2023).
European Geoparks Network. (2011). *Arouca declaration on geotourism*. www.europeangeoparks.org/?p=223 (accessed on 30 April 2024).
Folkersen, M., Fleming, C., Hasan, S. (2018). Deep-sea mining's future effects on Fiji's tourism industry: A contingent behaviour study. *Marine Policy*, 96, 81–89.
Ghazi, J., Hamdollahi, M., Moazzen, M. (2021). Geotourism of mining sites in Iran: An opportunity for sustainable rural development. *International Journal of Geoheritage and Parks*, 9(1), 129–142.
Hobbs, P.J., Mills, P.J. (2012). *Managing the threats to the Karst water resources of the cradle of humankind world heritage site, South Africa, Gauteng tourism authority*. www.gauteng.net/cradleofhumankind/management/reports_available/ (accessed on 12 June 2015).
Human Rights Watch. (2017). *Africa's natural resources: From curse to a blessing*. www.hrw.org/news/2017/04/21/africas-natural-resources-curse-blessing (accessed on September 2019).
Janneh, A., Ping, J. (2011). *Minerals and Africa's development*. Addis Ababa: United Nations Economic Commission for Africa, pp. 2–66.
Karombo, T. (2020). Zimbabwe has okayed a Chinese coal-mining pact at its top game reserve, even as elephants die. *Quartz Africa*, 3 September. https://qz.com/africa/1899381/zimbabwe-okay-chinese-coal-mining-in-hwange-park-as-elephants-die/ (accessed on 25 March 2021).
Knutsen, C.H., Kotsadam, A., Olsen, E.H., Wig, T. (2017). Mining and local corruption in Africa. *American Journal of Political Science*, 61(2), 320–334. https://doi.org/10.1111/ajps.12268.

Kumah, D., Nyarko, E. (2018). Gold mining and its effects through the lens of an archaeologist: Experiences from the Prestea Area, South-Western Ghana. *West African Journal of Applied Ecology*, 26(SI), 133–148.

Laker, C. (2023). Environmental impacts of gold mining – with special reference to South Africa. *Mining*, (2), 205–220.

Leonard, L. (2016). Mining and/or tourism development for job creation and sustainability in Dullstroom, Mpumalanga. *Local Economy*, 31(1–2), 249–263.

Leonard, L. (2017). State governance, participation and mining development. *Politikon*, 44(2), 327–345.

Leonard, L. (2019). Communities, mining corporations and corruption in South Africa. *The Conversation*, 7 February. https://theconversation.com/communities-mining-corporations-andcorruption-in-south-africa-111007 (accessed on 22 February 2019).

Leonard, L. (2023). Climate change, mining development and residential water security in the uMkhanyakude District Municipality, KwaZulu-Natal, South Africa: A double catastrophe for local communities. *Local Environment*, 28(3), 331–346. https://doi.org/10.1080/13549839.2022.2136644.

Leonard, L., Langton, A. (2016). Challenges facing tourist attractions due to acid mine drainage in the West Rand, Gauteng. *African Journal of Hospitality, Tourism and Leisure*, 5(1), 1–12.

Leonard, L., Lebogang, T. (2018). Exploring the impacts of mining on tourism growth and local sustainability. *Sustainable Development*, 26(3), 206–216.

Leonard, L., Lidskog, R. (2021). Examining reflexive scientisation and scientific expertise in South Durban, South Africa. *Journal of Risk Research*, 24(9), 1127–1140.

Magadza, C.H.D., Zinyama, L.M. (2007). Geotourism in Africa: The Zimbabwean case study. *African Journal of Business Management*, 1(5), 138–144.

Malijani, O. (2020). Mining in the context of African world heritage cultural landscapes: An assessment of management planning effectiveness. Doctoral thesis, Brandenburg University of Technology, Cottbus-Senftenberg.

Matoussi, H., Raissi, N., Abd Elmola, A., Laridhi, N. (2023). Does geology meet archaeology? New vision and opportunity for the development of geotourism: A case study of the Archaeo-Metamorphic-Geosite of Chemtou/Simitthus (Northwest Tunisia, North Africa). *Geoheritage*, 15(4). https://doi.org/10.1007/s12371-023-00888-9.

Matshusa, K., Leonard, L. (2023). Proposed governance model for geotourism and geoheritage in South Africa and the implications for Africa. *Geoheritage*, 15(1). https://doi.org/10.1007/s12371-022-00783-9.

Matshusa, K., Leonard, L., Thomas, P. (2021). Challenges of geotourism in South Africa. *Resources*, 10(11), 108.

McNutt, K., Rayner, J. (2010). Valuing Metaphor: A constructive account of reflexive governance in policy networks. In *Proceedings of the 5th international conference on interpretative policy analysis*, Grenoble, France, 23–25 June 2010.

Meißner, S. (2021). The impact of metal mining on global water stress and regional carrying capacities – a GIS-based water impact assessment. *Resources*, 10, 120. https://doi.org/10.3390/resources10120120.

Mir, R.A., Ahmad, S.A., Akaram, V., Kumar, V. (2023). Geoheritage and geotourism potential of Kalaruch old workings (ancient mines) and Satbarren (archeological stone) in Kashmir Valley of Himalaya. *Geoheritage*, 15(2). https://doi.org/10.1007/s12371-023-00807-y.

Mukwada, G., Sekhele, N. (2017). The potential of community-based geotourism in rural development in South Africa: The case of Witsie cave project. *Journal of Asian and African Studies*, 52(4), 471–483. https://doi.org/10.1177/0021909615595991.

Muma, D., Besa, B., Manchisi, J., Banda, W. (2020). Effects of mining operations on air and water quality in Mufulira district of Zambia: A case study of Kankoyo township. *Journal of the Southern African Institute of Mining and Metallurgy*, 120, 287–298.

Newsome, D., Dowling, R.K., Leung, Y.F. (2012). The nature and management of geotourism: A case study of two established iconic geotourism destinations. *Tourism Management Perspectives*, 2–3, 19–27.

Ngang, C., Kamga, S. (2022). *Natural resource sovereignty and the right to development in Africa*. New York: Routledge.

Norman, N. (2013). *Geological journeys: A traveller's guide to South Africa's rocks and landforms*. Cape Town: Penguin Random House.

Nyamekye, S., Mensah, P. (2012). Mining or our heritage? Indigenous local people's views on industrial waste of mines in Ghana. In: Show, K.-Y. (Ed.), *Industrial waste*. London: Intech, pp. 151–172.

Ochieng, G., Ephrahim, S., Onyeka, N., (2010). Impacts of Mining on Water Resources in South Africa. *Scientific Research and Essays*, 5(22), 3351–3357.

Ólafsdóttir, R., Tverijonaite, E. (2018). Geotourism: A systematic literature review. *Geosciences*, 8(7), 1–16.

Orr, R. (2002). Governing when chaos rules. *Washington Quarterly*, 25(4), 139–152.

Panchia, Y. (2023). Mining in South Africa: Employment threats and logistical challenges. *Mining Review Africa*, 23 December. www.miningreview.com/base-metals/south-africa-employment-threats-and-logistical-challenges/ (accessed on 15 February 2024).

Pearce, F. (2017). Murder in Pondoland, *The Guardian*, 28 March. www.theguardian.com/environment/2017/mar/27/murder-pondoland-how-proposed-mine-brought-conflict-southafrica-activist-sikhosiphi-rhadebe (accessed on 22 February 2019).

Poblete Piedrabuena, M.Á., Martí Molist, J., Beato Bergua, S. (2019). The deterioration of geoheritage in the central Spanish volcanic region by open-pit mining. *Geoheritage*, 11, 1903–1917. https://doi.org/10.1007/s12371-019-00405-x

Pretorius, M. (2016). Tourism tragedy unfolding on the Wild Coast. *Fin24*, 9 October.

Ruban, D. (2015). Geotourism – a geographical review of the literature. *Tourism Management Perspectives*, 15, 1–15.

Sallam, E.S., Ruban, D.A., Mostafa, M.T., Elkhodery, M.K., Alwilily, R.L., Molchanova, T.K., Zorina, S.O. (2020). Unique desert caves as a valuable geological resource: First detailed geological heritage assessment of the Sannur Cave, Egypt. *Arabian Journal of Geosciences*, 13(3). https://doi.org/10.1007/s12517-020-5176-4.

Scoon, R.N. (2020). Geotourism, iconic landforms and island-style speciation patterns in national parks of East Africa. *Geoheritage*, 12(3). https://doi.org/10.1007/s12371-020-00486-z.

Seccombe, A. (2018). Wild Coast mining case crystallises need to listen to communities. *Business Day*, 25 April.

Singh, R., Ghosh, P. (2019). Geotourism potential of coal mines: An appraisal of Sonepur-Bazari open cast project, India. *International Journal of Geoheritage and Parks*, 9(2), 172–181. https://doi.org/10.1016/j.ijgeop.2021.02.007.

Stirling, A. (2006). Precaution, foresight, sustainability: Reflection and reflexivity in the governance of science and technology. In: Voß, J.P., Bauknecht, D. & Kemp, R. (Eds.), *Reflexive governance for sustainable development*. Cheltenham, UK: Edward Elgar, pp. 225–272.

Taru, P., Chingombe, W. (2016). Geoheritage and the potential of geotourism in the Golden Gate Highlands National Park, South Africa. *African Journal of Hospitality, Tourism and Leisure*, 5(2), 1–11.

Transparency International. (2024). www.transparency.org/en (accessed on 1May2024).

United Nations Environment Programme. (2021). *Our work in Africa*. www.unep.org/regions/africa/our-work-africa (accessed on 1 June 2021).

Van Marissing, E. (2005). Citizen participation in the Netherlands: Motives to involve citizens in planning processes. In *ENHR conference "housing: New challenges and innovations in tomorrow's cities"*, Reykjavik, Iceland, 29 June–3 July.

Wesselink, A., Paavola, J., Fritsch, O., Renn, O. (2011). Rationales for public participation in environmental policy and governance: Practitioners' perspectives. *Environment and Planning A*, 43(11), 2688–2704.

Worlanyo, S., Jiangfeng, L. (2021). Evaluating the environmental and economic impact of mining for post-mined land restoration and land-use: A review. *Journal of Environmental Management*, 279, 111623. https://doi.org/10.1016/j.jenvman.2020.111623.
World Bank. (2019). *Digging beneath the surface: An exploration of the net benefits of mining in Southern Africa*. Washington, DC: World Bank. https://openknowledge.worldbank.org/bitstream/handle/10986/32107/Digging-Beneath-the-Surface-An-Exploration-of-the-Net-Benefits-of-Mining-in-Southern-Africa.pdf?sequence=1&isAllowed=y.
World Wildlife Fund (WWF). (2023). *Zambia cancels destructive mine in Lower Zambezi National Park*. https://wwf.panda.org/wwf_news/?9585966/Zambia-cancels-destructive-mine-in-Lower-Zambezi-National-Park (accessed on January 2024).

6 The potential contribution of geotourism in environmental protection and sustainable development in Cameroon

Henry Bikwibili Tantoh, Felix Ekwabena Donkor, and Tracey T. McKay

Introduction

Tourism is a multidimensional, multifaceted activity that touches many lives and many different economic activities (Cooper et al., 2008). People have always migrated for social, cultural, and economic reasons and personal or professional purposes. Additionally, globalisation has made tourism a popular global leisure activity. Recently, the notion of geotourism has been propagated as one of the latest concepts in the last few decades due to its rapid growth (Ólafsdóttir & Tverijonaite, 2018). Geotourism enhances and promotes the particularity of a region based on the complex connection between tourism and geology, heritage, culture, aesthetics, the health of the surrounding environment, and the population (European Geoparks Network, 2011). According to Hose (2008), geotourism refers to the provision of interpretive and service facilities to enable tourists to acquire knowledge and an understanding of the geology and geomorphology of a site (including its contribution to the development of Earth sciences) beyond a mere aesthetic appreciation. Geotourism has grown rapidly over the last few decades (Ólafsdóttir & Tverijonaite, 2018). It is a form of natural area tourism that specifically focuses on geology and landscape. Geotourism has been reported as one of the latest concepts in tourism studies that is environmentally innovative (Ólafsdóttir & Tverijonaite, 2018). It refers to tourism that enhances the geographical character of a place, including the environment, culture, aesthetics, heritage, and well-being of its residents. It has been reported that geotourism is a new form of tourism based on the geological environment (Dowling, 2013). The concept is rooted in the idea of promoting responsible and sustainable tourism that not only benefits the tourists but also contributes to the preservation and conservation of the environment. Thus, geotourism is the continuous effective utilisation of geoheritage sites for tourism activities to benefit local communities (Matshusa et al., 2021). Furthermore, geotourism is seen as a particular form of tourism whose main focus is geosites and which champions sustainable tourism (Dowling, 2013; Ngwira, 2015). Geotourism activities include hiking/leisure bush walks and rock climbing, which are subjects of this study.

Africa is endowed with abundant geotourism potential, such as national parks, biosphere reserves, and many other protected areas. It is argued that the appreciation

DOI: 10.4324/9781032706764-6

of geological resources is fundamental to geotourism (Rop, 2020). This is because geological resources foster an understanding of and connection with geological features through visitation, interpretation and education. The geodiversity of Africa includes the Doukkala-Abda and Tazenakht geopark in Morocco; Hoggar, Tassili, and Taza in Algeria; Brandberg, Dolerite Hills, Etendeka Plateau, and Namib Desert in Namibia; Table Mountain, Cradle of Humankind, and Witwatersrand gold fields in South Africa; Cheringoma Caves and columnar basalts of Lebombo Range in Mozambique; Fresco Escarpment in Côte d'Ivoire; Olumo batholiths, Jos Plateau in the central plateau area, and Ogbunike Caves in Nigeria; and the Mandara Mountains, Mount Fako, Lake Mwaneguba, and Lake Nyos in Cameroon (Errami et al., 2015; Rop, 2020). Africa's biodiversity and geodiversity, and significant geotourism potential, offer the prospect of cultural geotrails for both education and tourism (Rop, 2020). These geodiversity and biodiversity jointly form unique landscapes that are considered tourist attractions (Doumenge et al., 2020). The geological formations have long attracted visitors worldwide. These resources, therefore, require effective planning and appropriate development to avert adverse impacts of geotourism on geological features (Szepesi et al., 2016). However, these sites are at risk of irreparable destruction due to an increasing population, rising urbanisation, intensification of agriculture, and environmental degradation, among others. Furthermore, most of the communities with these potentials are relatively poor, such as is the case in the Korup National Park, the WAZA Park, and the Kruger National Park among others (Tefogoum et al., 2020; Matshusa et al., 2021). It is therefore imperative to conserve the environment because of the complex relation between tourism and the environment. The nature of geotourism calls for the use of geological landscapes that carry living things. It can therefore be argued that biodiversity is dependent upon geodiversity, since geoconservation deals with the conservation of the abiotic parts of the natural environment and plays a significant role in bioconservation and environmental protection (Rop, 2020).

Nestled in Central Africa, Cameroon boasts a rich tapestry of ecosystems, from dense rainforest to arid savannas, picturesque mountains, and a pristine coastline along the Gulf of Guinea. This remarkable biodiversity and geodiversity are not only a source of immense beauty but also vital for the well-being of both the planet and its people. However, these invaluable resources face numerous threats from deforestation, habitat destruction, increasing urbanisation, and the pressures from a growing population. Cameroon is renowned for its outstanding and abundant natural resources and natural beauty, with numerous remarkable geological features (Tefogoum et al., 2019). Despite the remarkable geoheritage sites in Cameroon, geoconservation has focused on the conservation of biodiversity. Geotourism and geoconservation are inadequately established to effectively respond to Cameroon's vision of economic emergence by 2035 and the United Nations Sustainable Development Goals (SDGs). The contribution of geotourism to Cameroon's emergence vision is silent in the country's vision 2035 agenda (Republic of Cameroon, 2009). Furthermore, the process of consuming tourism potentials adversely affects the environment (Wall & Mathieson, 2006). For example, excessive overuse of ecosystems due to pressure from visitors has magnified the chances of adverse environmental impacts and the pressure on the ecosystem as a result of overuse by visitors, thereby magnifying the chances of negative environmental impacts arising between the resource-protected areas and the products of geotourism (Rop, 2020; Harilal et al., 2021). Several studies have been conducted on tourism and ecotourism (Sharples, 2002; Hose, 2008), tourism principles and practice (Cooper et al., 2008), the typology of geotourism assets (Tefogoum et al., 2019), modelling the impact of geotourism on geoconservation (Rop, 2020), the contribution of geotourism to social sustainability (Matshusa et al., 2021), and geoheritage and geoparks (Global Geoparks Network, 2010; Errami et al., 2015), among others. However, research on the potential contribution of geotourism in environmental protection and sustainable development in Cameroon is limited. This chapter, therefore, investigates the potential contribution of geotourism in environmental protection and sustainable development in Cameroon. Acknowledging the interdependent rapport, relevance, and values of geotourism to geoconservation as a dynamic approach to preservation and maintenance of protected areas management is central to promoting ecosystem stewardship and contributing to the achievement of SDGs.

Geotourism and geoconservation: a literature review

Geotourism is one of the latest notions within tourism studies. The reputation of geotourism has quickly grown over the past few decades. It is argued that the growing popularity of geotourism initiates a dire need to investigate the potential contribution of geotourism in environmental protection and sustainable development (Ólafsdóttir & Tverijonaite, 2018). Noteworthy is that geotourism as an economic activity presents an opportunity to generate growth, employment, and conservation. Exceptional geological formations have long attracted visitors globally. In Iceland, for example, most travelogues written by foreign visitors in the eighteenth and nineteen centuries testified that the main reason of their visits was the country's geological formations. Similarly, geology-based tourism originated in Great Britain in the later part of the seventeenth century (Hose, 2008). It is possible that geotourism generally proliferated in other European countries in the same period.

The notion of geotourism has, therefore, been redefined in the past two decades by different scholars from different continents and countries. However, the definitions are not similar. For example, scholars in Australia and United Kingdom classify geology and landscape as central elements of geotourism (Joyce, 2007) while those in the United States use a broader approach to the definition of geotourism, referring to it as "tourism that sustains or enhances the geographical character of the place being visited, including its environment, culture, aesthetics, heritage and the well-being of its residents" (Stokes et al., 2023). Around the same time, geotourism was being defined as a new form of tourism, which is "a multi-interest kind of tourism exploiting natural sites and landscapes containing interesting earth-science features in a didactic and entertaining way" (Pralong, 2006). This definition, however, reveals that geotourism is a subdivision of sustainable tourism that, instead of focusing on reducing the impact on the ecological environment, seeks to preserve all natural and human attributes that render a given location distinct from others (Ólafsdóttir & Tverijonaite, 2018). Subsequent partnerships between National Geographic and local organisations led to the publication of Geotourism MapGuides which exhibit about 20 tourism destinations globally that enlighten tourists regarding more sustainable sites in the area (Bosak et al., 2010). This has greatly enhanced the geographical strength of the region and contributed to the livelihood and well-being of those in close proximity to the resources. It has been reported that the National Geographic publications have greatly proliferated the idea of geotourism as a sustainable approach to tourism development. However, the importance of narrowing the definition of geotourism to geology and conservation is critical, because applying a broader approach to geotourism might reduce the impact of the concept (Hose, 2016). Others have argued that a more specific definition of geotourism would advance a more decisive approach to the concept (Newsome & Dowling, 2010). This is essential for attaining the broader aims of geotourism, such as geodiversity conservation, visitor education, and empowering local communities by providing knowledge about their geological resources and employment opportunities (Newsome & Dowling, 2010). Furthermore, it contributes to the integration of protected areas.

Several studies have documented that the growth in tourism can make positive contributions to the sustainability of protected areas and the development of the surrounding communities (Dowling, 2013; Štrba et al., 2020). Thus, it has commonly been assumed that geotourism is sustainable tourism with a primary focus on experiencing the Earth's geological features in a way that fosters an environmental and cultural understanding, appreciation and conservation, and is locally beneficial (Dowling, 2013). It is about creating a geotourism product that protects geoheritage, helps to build communities, communicates and promotes geological heritage, and works with a wide range of different people (Dowling, 2013; Ngwira & Kankhuni, 2018; Matshusa et al., 2021). Tourism has, therefore, become the main instrument not only for preserving these areas but also for raising awareness of the environment of the population and visitors. Geotourism, a concept that integrates sustainable tourism with the preservation of natural and cultural heritage, holds immense potential for contributing to environmental protection and sustainable development (Dowling, 2013).

Studies suggest that greater attention should be given to environmentally innovative forms of tourism that foster both environmental and social responsibility (Pásková, 2012; Harilal et al., 2021). Thus, a new emerging type of tourism that is "environmentally innovative" is referred to as geotourism. In view of the importance of geotourism, geological attractions are being developed around the world, mostly as a sustainable development tool for the development of local and regional communities. This is because geotourism generates economic opportunities for local communities. In addition, sustainable tourism practices support the livelihoods of indigenous populations by creating jobs, encouraging the development of ecofriendly accommodation and promoting local craftsmanship and cuisine (Harilal et al., 2021). Economic empowerment, in turn, motivates communities to protect their natural environment, recognising its vital role in attracting visitors and sustaining their way of life. Given that, geotourism seeks to conserve natural and cultural resources, provide economic benefits, particularly to the population closest to the geologic site, create awareness for both tourists and the locals, and can be a powerful tool for achieving the SDGs.

Despite the abundant geotourist attractions in Africa and Cameroon, tourism facilities are inadequately developed, valued or professional. With the exception of a few countries like Rwanda and São Tomé and Príncipe, tourism remains a neglected sector in most national economies. It has been argued that tourism development policies and strategies are just beginning to be formulated, and governments do not appear to be giving the attention needed to their implementation (Viard, 2008; Kimbu & Tichaawa, 2018). This activity continues to face many obstacles, including red tape, complex visa requirements, poor infrastructure, insecurity and armed conflict, lack of specialised skills, lack of investment, and an unfavourable business climate (Kamga Kamdem et al., 2020). Tourism therefore could play an important role in biodiversity protection, providing opportunities for the socioeconomic development of protected areas and economic justification for the protection of species and ecosystems (WTTC, 2020; Harilal et al., 2021).

Materials and methods

Description of the study site

Cameroon, a diverse and ecologically rich nation in Central Africa, is endowed with a remarkable blend of natural wonders, from dense rainforest to awe-inspiring mountain ranges and an array of unique wildlife (Tefogoum et al., 2022). Cameroon boasts a rich tapestry of biodiversity, dramatic landscapes, and a diverse cultural heritage (see Figure 6.1). Leveraging these assets through geotourism can foster sustainable economic growth while simultaneously safeguarding the environment. Cameroon's unique topography encompasses everything from the breathtaking beaches along the Gulf of Guinea to the towering peaks of the Cameroon Volcanic Line and the dense forest of the Congo Basin. These natural platforms for geotourism initiatives highlight the country's geological, ecological, and cultural diversity. As the world grapples with pressing environmental concerns and seeks sustainable development solutions, geotourism emerges as a promising avenue for addressing both ecological preservation and economic growth in Cameroon (Harilal et al., 2021). Cameroon's geotourism potential is abundant with attractions such as Mount Cameroon (commonly called Mount Fako), the highest peak in Central and West Africa, unique volcanic lakes in the Western Highlands, and national parks such as Korup National Park and the WAZA Park. These natural wonders are integral to the country's identity and are essential for environmental conservation efforts.

Data collection

This desk study was conducted from March 2023 to February 2024. The initial stage of the study involved purposive selection to come up with a reasonable conclusion.

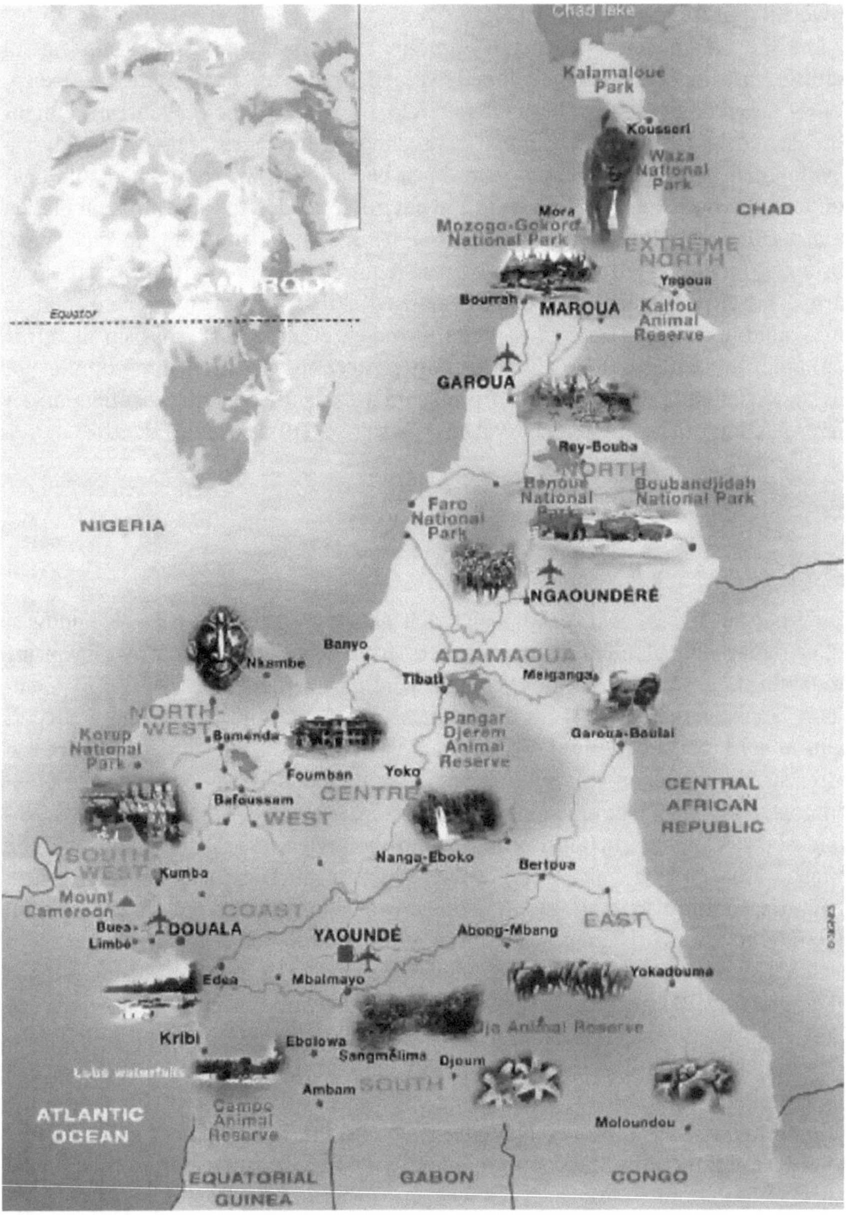

Figure 6.1 Map of Cameroon showing tourist potential.

Source: www.google.com/search?q=Touristic+map+of+Cameroon, 2024

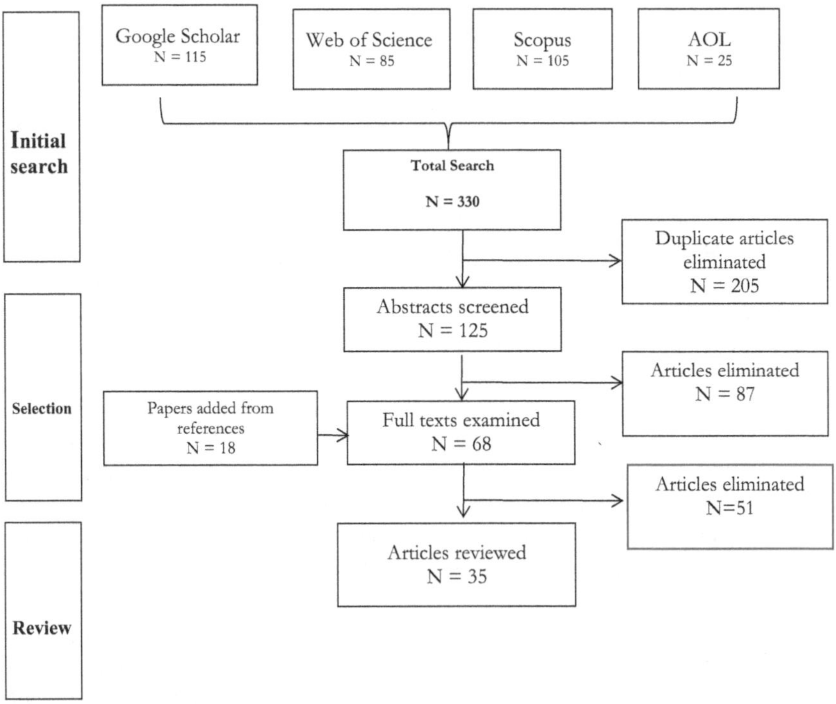

Figure 6.2 Flow chart showing the literature search and selection process.

This was to select only articles directly related to the study for analysis. The articles were selected from internet databases of indexed journals and other peer-reviewed publications such as Scopus, Web of Science, Google Scholar and AOL. From these sites, keywords like "geotourism", "geoconservation", "geoheritage", "environment", and "sustainable development" were typed in the search engine using a four-stage process (see Figure 6.2). This resulted in 330 peer-reviewed articles with data related to the study. In the second stage, the abstracts of the articles were screened and duplicates were filtered, resulting in 205 articles being eliminated.

Furthermore, about 87 articles were appraised, because they were relatively recent, to form the contextual and theoretical framework of this study. These articles were directly related to geotourism, geoconservation, and sustainable development. A further 51 studies that were not precisely related to geotourism, geoconservation and sustainable development were eliminated from the review. By examining the bibliographies of the remaining articles, an additional 18 related articles were identified. After assessing the complete articles that were added to the sample of articles, 35 research articles were selected for the review. A summary and synthesis of the most important results enabled the identification of the main research trends in geotourism research and of areas that require further research.

Results and discussion

Geotourism and geoconservation

Geotourism is a recent phenomenon based on geology, geomorphology, and landscapes (Boškov et al., 2015). It is a form of sustainable tourism that emphasizes the appreciation and conservation of the Earth's geological features and landscapes. It involves the promotion of responsible travel to geological and natural heritage sites. Thus, the concept has become a global phenomenon. Studies reveal that geology and landscapes are everywhere, and this has shaped their relevance as geotourism attractions (Nazaruddin, 2019; Štrba et al., 2020). The European Geoparks Network (2011) notes that geotourism enhances and promotes the particularity of a region based on the complex connection between tourism and geology, heritage, culture, aesthetics, the health of the surrounding environment, and the population. Thus, it presents an opportunity to address environmental challenges by promoting responsible tourism practices that respect the environment. This is because visitors explore national parks and protected areas, witness their diverse wildlife, and engage with local communities while contributing to conservation efforts (Nazaruddin, 2019). The nature of geotourism calls for the use of geological landscapes that carry living things. It is, therefore, suitable to state that geoconservation deals with the conservation of the abiotic parts of the natural environment and plays a significant role in bioconservation. This is because biodiversity is supported by geodiversity; and geotourism should be able to play an important role in preventing the degradation of landforms, soil and water that adversely impacts the biological species and surrounding communities (Sharples, 2002). Similarly, Dowling (2013) highlights that knowledge of geology is essential for a full understanding of the geological processes in a given region. Thus, geotourism follows the notable ABC approach, signifying that the abiotic components (geology and climate) determine the biotic components (flora and fauna), and combined these define the cultural elements, such as the lifestyle of people (Dowling, 2013). Acknowledging the symbiotic relationship, values, and relevance of geotourism to geoconservation as a dynamic approach to the preservation and maintenance of protected areas management is central to promoting ecosystem stewardship and contributing to the achievement of SDGs.

Environmental conservation through tourism

Recent evidence suggests that greater attention should be paid to environmentally innovative forms of tourism that foster both environmental and social responsibility (Pásková, 2012). In this regard, geotourism raises awareness about the importance of environmental protection, encouraging local and international stakeholders to support sustainable practices and policies (Neto & Henriques, 2022). National parks and protected landscape areas, for example, are crucial zones for implementing geoconservation strategies. This is due to their size and importance. Given that many protected areas have become important tourist destinations, they are suitable for raising public awareness of all aspects of nature, biological and geological

(Ruban, 2015). Geotourism and environmental protection play essential roles in promoting sustainable development by balancing the economic benefits of tourism with the preservation of natural and cultural resources. An implication of this is that geotourism has the potential to drive economic development. For example, by creating jobs, generating revenue, and fostering small-scale enterprises, it can reduce poverty and enhance the quality of life for local communities. This, in turn, can reduce the reliance on activities that harm the environment, such as illegal logging and poaching, while offering alternative and more sustainable sources of income. Thus, protected areas are already recognised by local authorities and people who live and work in such locations. Correspondingly, geotourism is sustainable tourism with a primary focus on experiencing the Earth's geological features in ways that foster an environmental and cultural understanding, appreciation of conservation, and is locally beneficial (Dowling, 2013).

Geotourism and sustainable development

Sustainable geotourism development typically involves minimising the negative environmental impacts of tourism. This includes responsible waste management, limiting the construction of infrastructure and promoting ecofriendly transportation options. Sustainable practices can also extend to accommodation, encouraging the use of renewable energy sources and water conservation. An important tool that facilitates sustainable geotourism development is the use of geoparks. These are "geographical areas where geological heritage sites are part of a holistic concept of protection, education and sustainable development" (Global Geoparks Network, 2010). Geotourism, a concept that integrates sustainable tourism with the preservation of the natural and cultural heritage, holds immense potential for contributing to environment protection and sustainable development. Thus, Cameroon (a country endowed with breathtaking natural beauty and ecological diversity) stands at the crossroads of an opportunity for sustainable development through geotourism and environmental protection. Geotourism (a concept that combines the appreciation of natural landscapes, cultural heritage, and responsible travel) has the potential to make a significant contribution to both the preservation of Cameroon's unique environment and the promotion of sustainable economic growth. Geotourism has the potential to be a win-win for Cameroon. This is because, it can stimulate economic growth, provide jobs, and promote the preservation of the country's rich natural and cultural heritage. This is in line with the observations of the World Travel and Tourism Council (WTTC, 2020) emphasising the importance of the tourism industry in job creation, accounting for 10.3 per cent of global gross domestic product in 2019. It contributes significantly to the creation of jobs in some countries, especially for women, youth, migrant workers, rural communities, and indigenous peoples (Kamga Kamdem et al., 2020). Furthermore, it has the potential to drive other economic sectors such as agriculture, transport, construction and telecommunications (WTTC, 2020). Regarding conservation, geotourism is an essential mode to both integrate protected areas into local and global economies and generate funding to cover, at least in part, their management costs (WTTC, 2010). By carefully planning and managing geotourism initiatives with a strong focus on sustainability, Cameroon can use this industry as a force for environmental protection and long-term sustainable development.

Conclusion

Geotourism, a concept that integrates sustainable tourism with the preservation of natural and cultural heritage, holds immense potential for contributing to environmental protection and sustainable development. This chapter investigated geoheritage sites in Cameroon and their contribution to geoconservation. Generally, Cameroon has the potential to contribute significantly to both environmental protection and sustainable development. By promoting responsible tourism practices, fostering a sense of environmental stewardship and offering economic benefits to local communities, geotourism can help to strike a balance between preserving the country's ecological treasures and fostering economic growth. As the country strives for a sustainable future, geotourism offers a path forward that respects the environment while embracing the potential of responsible tourism. It presents an opportunity to address environmental challenges by promoting responsible tourism practices that respect the environment. Geotourism, therefore, raises awareness about the importance of environmental protection, encouraging local and international stakeholders to support sustainable practices and policies. Furthermore, geotourism has the potential to drive economic development. By creating jobs, generating revenue and fostering small-scale enterprises, it can reduce poverty and enhance the quality of life for local communities. This, in turn, can reduce reliance on activities that harm the environment, such as illegal logging and poaching, while offering alternative and more sustainable sources of income.

In conclusion, geotourism can be a powerful driver for sustainable development by intertwining economic interests with the protection of the environment. However, it is crucial that geotourism is managed responsibly, with a long-term perspective in mind, to ensure that the natural and cultural treasures that attract tourists today will be preserved for future generations. Collaboration among governments, local communities, tourists, and environmental organisations is essential for achieving this delicate balance between geotourism and environmental protection. In addition, the integration of geotourism and environmental protection can offer a promising path towards sustainable development in Cameroon. In the same vein, geotourism development should be encouraged at all levels. Geotourism development at the local and regional levels, for example, must be developed within the context of sustainable local, national, and international tourism development. At the local, regional, and national levels, development policies, plans and programs, laws, decrees and regulations, and marketing, all influence sustainable tourism development. However, the three focal philosophies of sustainable development which can also be applied to regional geotourism development planning are its concentration on socio-economic and ecological concerns. Thus, geotourism can and will only be sustainable where there are benefits for those closest to the resources (host community); these may be social and/or cultural and environmental and might not necessarily be limited to economic benefits.

Recommendations

- Increased local investments. The ministry of tourism through its regional offices should encourage formalised businesses in the tourism value chain to increase employment opportunities and enhance economic activities.
- To public administrators and promoters of tourism: Given that more visitors are attracted to protected areas, there is a high probability of environmental degradation. Thus, pro-environmental conduct of geotourists in protected areas is encouraged.
- The concept of geotourism and geoparks are units of tourism and geology and are relatively new. Although still in the infancy stage and emerging as tourism niche markets awaiting further development and commercialisation, geotourism and geoparks needs to be encouraged by the government and related stakeholders and endorsed as instruments for sustainable tourism with permits to contribute to sustainable tourism development, rural development, local community participation and poverty alleviation.
- Improvement in the tourist industry generally would increase revenue which will in return contribute to re-establish notable localities, thereby encouraging local craftspeople. It can help to retain traditional cultural celebrations and performing arts that would otherwise disappear.

Acknowledgements

The financial assistance of the National Institute for the Humanities and Social Sciences-Council for the Development of Social Science Research in Africa (NIHSS-CODESRIA) for this research is hereby acknowledged. The opinions expressed and conclusions arrived at are those of the authors and are not necessarily to be attributed to the NIHSS-CODESRIA.

References

Bosak, K., Boley, B., Zaret, K. (2010). Deconstructing the "crown of the continent": Power, politics and the process of creating National Geographic's Geotourism MapGuides. *Tourism Geographies*, 12, 460–480.

Boškov, J., Kotrla, S., Jovanović, M., Tomić, N., Lukić, T., Rvović, I. (2015). Application of the preliminary geosite assessment model (GAM): The case of the Bela Crkva municipality (Vojvodina, North Serbia). *Geographica Pannonica*, 19(3), 146–152. https://doi.org/10.5937/GeoPan1503146B.

Cooper, C., Fletcher, J., Gilbert, D., Fyall, A., Wanhill, S. (2008). *Tourism: Principles and practice*. London: Pearson Education.

Doumenge, C., Michel, B., Poliwa, R., Breuer, T., De Chassey, E., Ngoga, T., Arranz, L. (2020). *Ecotourism and protected areas in Central Africa: A future in common*. file:///Users/a1/Desktop/Book%20Chapters/Book%20Chapter-%20Geotourism/EdAP_2020_ch8_EN.pdf (accessed on 21 February 2024).

Dowling, R.K. (2013). Global geotourism – an emerging form of sustainable tourism. *Czech Journal of Tourism*, 2(2), 59–79. https://doi.org/10.2478/cjot-2013-0004.

Errami, E. et al. (2015). Geoheritage and geoparks in Africa and the Middle-East: Challenges and perspectives. In E. Errami et al. (eds.), *From geoheritage to geoparks, geoheritage* (pp. 1–23). Geoparks and Geotourism. https://doi.org/10.1007/978-3-319-10708-0.

European Geoparks Network. (2011). www.europeangeoparks.org/?page_id=962 (consulted on 13 February 2024).

Global Geoparks Network. (2010). *Guidelines and criteria for national geoparks seeking UNESCO's assistance to join the global geoparks network (GGN)*. www.unesco.org/new/fileadmin/MULTIMEDIA/HQ/SC/pdf/sc_geoparcs_2010guidelines.pdf (accessed on 10 February 2024).

Harilal, V., Tichaawa, T.M., Saarinen, J. (2021). The impacts of ecotourism and conservation measures in protected areas on local communities in Cameroon. *Tourism Review International*, 25(2–3), 89–103.

Hose, T.A. (2008). Towards a history of geotourism: Definitions, antecedents and the future. *The Geological Society of London Special Publication*, 300, 37–60.

Hose, T.A. (2016). *Geoheritage and Geotourism: European Perspective*. Woodbridge: The Boyden Press.

Joyce, B. (2007). Geotourism, geosites and geoparks: Working together in Australia. *The Australian Geologist*, 144, 26–29.

Kamga Kamdem, S.L., Ngar-Odjilo, M., Njamen, B., Piga, B., Poliwa, R.C., Tabapssi, T. (2020). *Le Covid-19: Impacts sur le tourisme en Afrique centrale et mesures de relance*. Yaoundé, Cameroun: RETAC.

Kimbu, A.N., Tichaawa, M.T. (2018). Sustainable development goals and socio-economic development through tourism in Central Africa: Myth or reality? *GeoJournal of Tourism and Geosites*, 23(3), 780–796.

Matshusa, K., Leonard, L., Thomas, P. (2021). The contribution of geotourism to social sustainability: Missed opportunity? *The International Journal of Sustainability in Economic, Social and Cultural Context*, 17(1), 95–118. https://doi.org/10.18848/2325-1115/CGP/v17i01/95-118.

Nazaruddin, D.A. (2019). Selected geosites for geoheritage, geotourism and geoconservation in Songkhla province, Southern Thailand. *Quaestiones Geographic*, 38(1), 161–177.

Neto, K., Henriques, M.H. (2022). Geoconservation in Africa: State of the art and future challenges. *Gondwana Research*, 110, 107–113.

Newsome, D., Dowling, R.K. (2010). Setting an agenda for geotourism. In D. Newsome & R. Dowling (eds.), *Geotourism: The tourism of geology and landscape*. Oxford: Goodfellow Publishers Limited, pp. 1–12.

Ngwira, C., Kankhuni, Z. (2018). What attracts tourists to destinations? *African Journal of Hospitality, Tourism and Leisure*, 7(1), 1–19.

Ngwira, P.M. (2015). Geotourism and geoparks: Africa's current prospects for sustainable rural development and poverty alleviation. In E. Errami, M. Brocx & V. Semeniuk (eds.), *From geoheritage to geoparks, case studies from Africa and beyond.* Springer International Publishing, pp. 25–33. https://doi.org/10.1007/978-3-319-10708-0_2.

Ólafsdóttir, R., Tverijonaite, E. (2018). Geotourism: A systematic literature review. *Geosciences*, 8, 234. https://doi.org/10.3390/geosciences8070234.

Pásková, M. (2012). Environmentalistika cestovního ruchu (tourism environmentalism). *Czech Journal of Tourism*, 1(2), 77–113.

Pralong, J. (2006). Geotourism: A new form of tourism utilizing natural landscapes and based on imagination and emotion. *Tourism Review*, 61(3), 20–25. https://doi.org/10.1108/eb058476.

Republic of Cameroon. (2009). *Cameroun vision 2035.* www.cameroon-embassy.nl/wpcontent/uploads/2022/11/Cameroon_VISION_2035_English_Version.pdf (accessed on 20 March 2024).

Rop, W.C. (2020). Modelling the impact of geo-tourism on geo-conservation of Hell's Gate National Park in Kenya. *Ecofeminism and Climate Change*, 2(1), 17–25. https://doi.org/10.1108/EFCC-05-2020-0015.

Ruban, D.A. (2015). Geotourism – a geographical review of the literature. *Tourism Management Perspectives*, 15, 1–15.

Sharples, C. (2002). *Concepts and principles of geoconservation; Tasmanian parks and wildlife services*. https://parks.tas.gov.au/ (consulted 18 March 2024).

Stokes, A.M., Cook, S.D., Drew, D. (2023). *Geotourism: The new trend in travel*. Washington, DC: Travel Industry America and National Geographic Traveler.

Štrba, L., Kolaˇckovská, J., Kudelas, D., Kršák, B., Sidor, C. (2020). Geoheritage and geotourism contribution to tourism development in protected areas of Slovakia – theoretical considerations. *Sustainability*, 12, 2979. https://doi.org/10.3390/su12072979.

Szepesi, J., Harangi, S., Esik, Z., Novak, T.J., Lukacs, R., Soos, I. (2016). Volcanic heritage and geotourism perspectives in Hungary: A case of a UNESCO world heritage site, Tokaj wine region historic cultural landscape, Hungary. *Geoheritage*, 9(3), 329–349. https://doi.org/10.1007/s1237- 016-0205-0.

Tefogoum, G.Z., Dongmo, A.K., Nkouathio, D.G., Dedzo, M.G. (2022). The geodiversity of Lefo and Santa-Mbu Calderas (Bamenda Mountains, Cameroon Volcanic Line): Factor for socioeconomic activities. *International Journal of Geoheritage and Parks*, 10(4), 491–506.

Tefogoum, G.Z., Nkouathio, D.G., Dongmo, A.K., Dedzo, M.G. (2019). Typology of geotouristic assets along the south continental branch of the Cameroon Volcanic Line: Case of the Mount Bambouto's caldera. *International Journal of Geoheritage and Parks*, 7, 111–128.

Tefogoum, G.Z., Román, A.Q., Umaña, D.P. (2020). Geomorphosites inventory in the Eboga Volcano (Cameroon): Contribution for geotourism promotion. *Géomorphologie: Relief, Processus, Environnement*, 26(1), 19–33.

Viard, E. (2008). *Afrique centrale et tourisme. Reflexions autour du developpement touristique*. Libreville: RAPAC.

Wall, G., Mathieson, A. (2006). *Tourism: Change, impacts and opportunities*. England: Pearson Education Limited.

World Travel and Tourism Council (WTTC). (2010). *The 2010 travel and tourism economic research. Economic data research tool*. World Travel and Tourism Council. www.wttc.org.

World Travel and Tourism Council (WTTC). (2020). *Travel and tourism. Global economic impact and trends*. London: WTTC.

7 Mount Bambouto slopes geomorphosites (Cameroon Volcanic Line)

A geological heritage for geotouristic and geoeducational practices

Ghislain Zangmo Tefogoum, Merlin Gountié Dedzo, Hyacinthe Zouyane Nouhou Dama, Raoul Merlin Ndonbou, Armand Kagou Dongmo, and David Guimolaire Nkouathio

Introduction and geological setting

Volcanic events occurred in Cameroon for several million years. Numerous volcanoes were built up and spread in a N30°E direction. This set of volcanoes has been named the Cameroon Volcanic Line (Figure 6.1). It is an intraplate volcanic province located on the western edge of the African continent at the equator, extending over both the Atlantic Oceanic lithosphere and the African continental lithosphere (Tabod et al., 1992; Déruelle, 1994). This chain of intraplate volcanoes is over 1,600 km long and about 100 km wide, stretching from the Island of Pagalu in the Atlantic Ocean to Lake Chad (Tchoua, 1974; Burke, 2001). The continental domain, which is approximately 900 km long and dominates Cameroonian territory, is a succession of horsts and grabens (Nkouathio et al., 2008). The horsts are made of large polygenetic volcanoes or volcanic plateaus, characterised by complete magmas series, whereas the grabens are monogenetic volcanic fields displaying basic magmas suites (basanites, basalts, and accessory hawaiites) (Tamen et al., 2007).

The origin of the Cameroon Volcanic Line has been controversial up until very recently, despite numerous studies. According to Nana's (2001) findings, the LVC is an incipient rift. According to Halliday et al. (1988), it is the result of one or more hot spots. Finally, Ngako et al. (2006) and Njonfang et al. (2011) conclude that the LVC is the result of a complex interaction between hotspots and Precambrian faults during the displacement of the African plate. Mount Bambouto is the third largest volcano (by volume) in the Cameroon Volcanic Line, after Mount Cameroon and Mount Manengouba. The Mount Bambouto slopes lie in the north-east extension of the Mount Manengouba, from where they are separated by the Mbô plain; they adjoin Mount Bambouto almost continuously to the north-east and cover an area of around 800 km^2. They lie between latitudes 05°25' and 05°50'N and longitudes

DOI: 10.4324/9781032706764-7

09°55' and 10°15'E. They straddle the departments of Bambouto to the east, Menoua to the south, Lebialem to the west, and Mezam to the north-west, with Mount Mélétan rising to 2,744 m (Zangmo Tefogoum et al., 2019, 2020a). Mount Bambouto is an enormous shield volcano with a general south-west–north-east orientation (Tchoua, 1972). This massif is split open at the summit by a larger caldera, located between 05°37'–05°46'N and 09°56–10°07'E.

(Zangmo Tefogoum et al., 2019). The Mount Bambouto caldera and slopes are constituted by mafic (less) and felsic rocks (more). The mafic rocks are characterised by tephrites, basanites, alkaline olivine basalts, hawaiites, and mugéarites (Youmen et al., 2005; Marzoli et al., 2000; Nkouathio et al., 2008). Felsic lavas are more widespread than mafic ones, accounting for around two-third of the volcanic rocks and are made up of trachytes, rhyolites, phonolites, ignimbrites, tuffs, and, secondarily, benmoreites (Kagou Dongmo et al., 2010; Gountié Dedzo et al., 2020) (Figure 7.1). These rocks crop out in flows, domes, and dykes of various altitudes, forms, and dimensions that underpin the regional topography.

Alteration of these rocks produces several types of soil. Young soils of andosolic character (Tematio et al., 2011) that are black, light grey, light brown to reddish brown in colour, developed on trachytes, phonolites, and tuffs. On the outer slopes of the caldera, ferrallitic andic intergrades are found on younger volcanic rocks (Tematio et al., 2004). In places, these soils cover the flows, flow domes, intrusion domes, and dykes that constitute the main outcrops of volcanic rocks in the study area. In the Mount Bambouto region, sedimentary rocks are generally represented by clays, lignite clays, and sandstones of various facies that outcrop in small limnic basins such as Dschang (Kenfack et al., 2011).

A set of names has been proposed in the literature by many authors to classify landforms. Geotope and geosite are considered by Bussard (2014) as synonyms and they are used by German-speaking, English-speaking, and Italian-speaking authors. Several definitions of geotope have been proposed (Panizza & Piacente, 1993; Grandgirard, 1995; Panizza, 2001; Panizza & Piacente, 2003; Reynard, 2004). Reynard (2004) defines a geotope as a geological or geomorphological object that has acquired a scientific (e.g. sedimentological stratotype, relict moraine representative of a glacier extension), cultural/historical (e.g. religious or mystical value), aesthetic (e.g. some mountainous or coastal landscapes), and/or social/economic (e.g. aesthetic landscape as tourist destination) value due to human perception or exploitation. In his typology of geotopes, Grandgirard (1997) established a set of 10 types of geotopes. The type that highlights geological landforms is called geomorphological geotopes. Since this study focuses on geomorphological units on Mount Bambouto slopes, this type has been chosen.

Over the years, geomorphological geotopes have been given numerous designations (Reynard, 2004): geomorphological assets (Panizza & Piacente, 1993), *biens géomorphologiques* (Carton et al., 1994), geomorphological sites (Hooke, 1994), *géotopes géomorphologiques* (Grandgirard, 1995; Grandgirard, 1997), sites of geomorphological interest (Rivas et al., 1997), and geomorphosites (Panizza, 2001). For this work, we retain the term 'geomorphosites', which according to

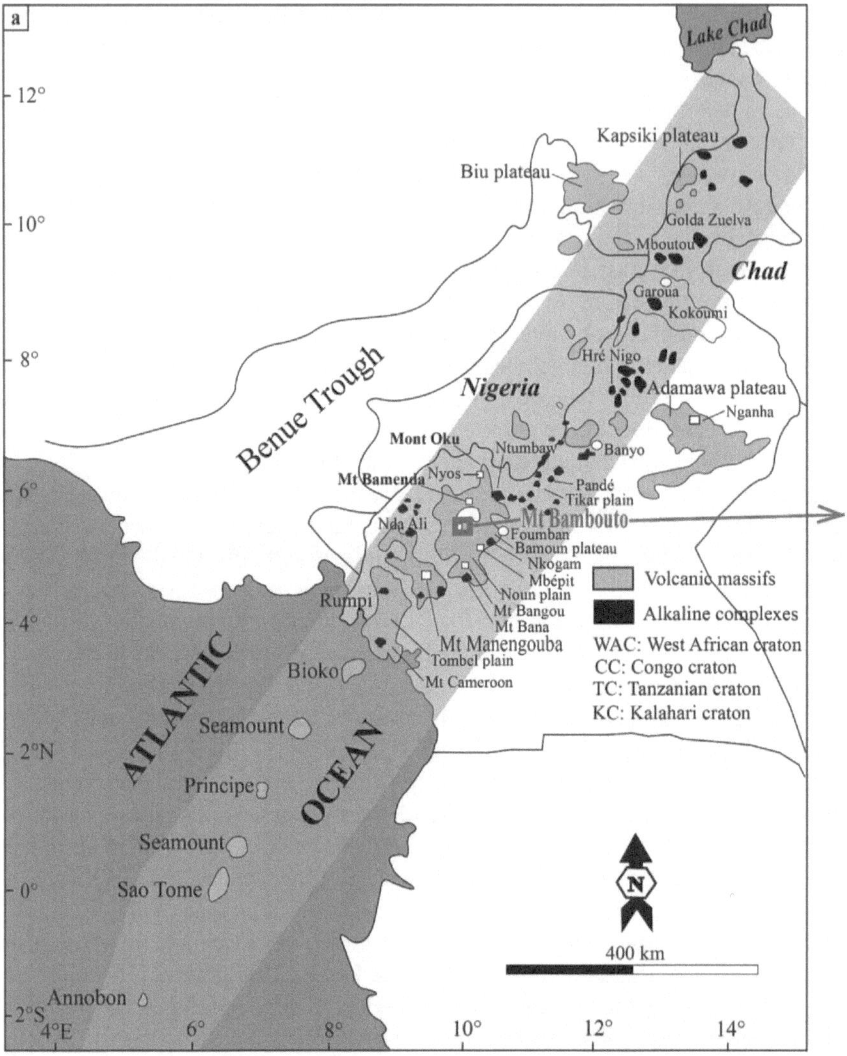

Figure 7.1 Location of the study area and Cameroon Volcanic Line and Mount Bambouto.

Sources: Burke, 2001; Kagou Dongmo et al., 2010

Bussard (2014) is a contraction of 'geomorphological' and 'sites'. According to Panizza (2001), geomorphosites are landforms that have acquired a scenic/aesthetic, scientific, cultural, historical, and/or a social/economic value due to their human perception or exploitation. For Shayan et al. (2011), geomorphosites are also defined as the geomorphological landforms that have gained scientific, historical, cultural, aesthetic, or socioeconomic value due to human knowledge and exploitation. The concept of geomorphosites has expanded rapidly in the last few

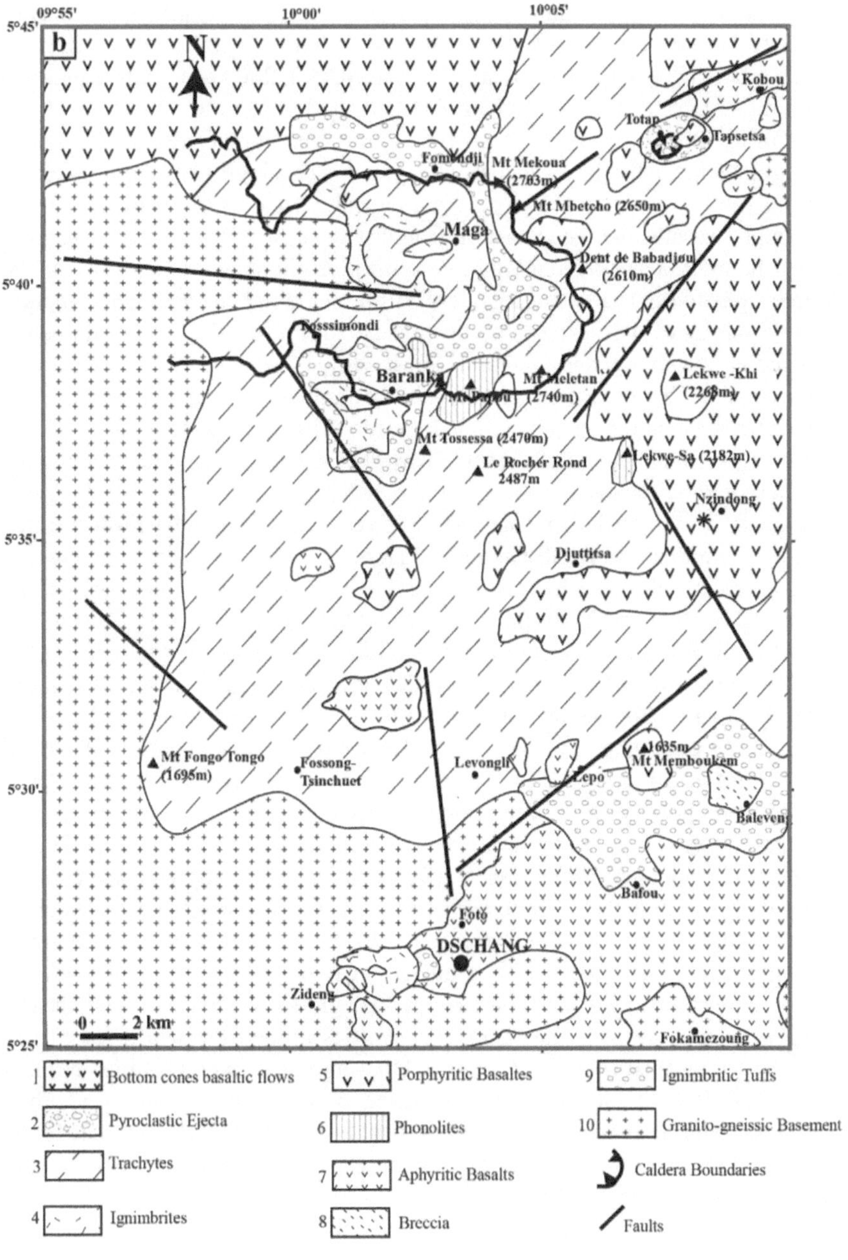

Figure 7.1 (Continued)

decades. In most cases, geomorphosites are one of the paramount components of the United Nations Educational, Scientific and Cultural Organization (UNESCO) Global Geopark (UGGp) label created by UNESCO in 2004 and the International Union of Geological Sciences (IUGS) Geological Heritage Sites label created by the IUGS in 2022. In many regions of the world, geomorphosites are threatened by anthropogenic activities or natural phenomena. This highlights the problem of protecting and conserving the world's geological heritage sites, which generally have a high tourist value. The value of geomorphosites is very important considering their future protection and touristic capitalisation (Comanescu et al., 2011). The relationship between tourism and geological sites and their characteristics (including the diversity of landscapes) can be considered a relatively new phenomenon and a subset of geology and tourism (Dincă, 2007). Therefore, the assessment of geomorphosites should be considered a tool for nature conservation (geoconservation), natural resource management, and the formulation of geotourism strategies in the area (Krishnanand, 2019). That is why UGGp plays a paramount role in the protection of geomorphosites.

Although Cameroon does not have a UGGp yet, it does have an IUGS-designated site, one of the first 100 IUGS Geological Heritage Sites. This site is the Quaternary Cameroon Volcano, which is the most active volcano in Cameroon, having last erupted in 1999 and 2000 (Suh et al., 2003; Wantim et al., 2013). This clearly shows that the volcanic regions of Cameroon have an exceptional geological heritage. These are undeniable assets for the development of geotourism, as demonstrated by Zangmo Tefogoum et al. (2014, 2017, 2020b) and Ziem à Bidias et al. (2020, 2023). According to Newsome and Dowling (2010), geotourism is defined as a form of tourism practised in natural spaces that specifically focuses on landscape and geology and their promotion, interpretation, and conservation through appreciation and education. Geotourism is a form of tourism that has experienced significant development in the last 20 years (Emil, 2020). Applying the concept of geotourism to particular landforms is the best way to transfer geomorphological knowledge to society (Zgłobicki et al., 2015). It is noteworthy that natural and anthropogenic landforms in the study area can be relevant assets for sustainable geotourism activities. Geotourism is a new global phenomenon that attracts several million tourists a year in many countries around the world, especially those with established geoparks (Lopes & Diawara, 2023). For example, in 2020, Dunhuang UNESCO Global Geopark received 6,5844 million tourists, yielding 8.03 billion yuan in total tourism revenue (Cai et al., 2023). The different forms of tourism in Cameroon are cultural tourism, ecotourism, religious tourism, and historical tourism (Zangmo Tefogoum et al., 2022a).

Geotourism is far from being included in the diversification policy for this economic sector. Although the geological heritage is quite diverse and innumerable, it undoubtedly suffers from a lack of knowledge on the part of decision-makers and the general public. This study is a continuation of that carried out by Zangmo Tefogoum et al. (2019) in the Mount Bambouto caldera, focusing on the slopes of the massif to carry out an inventory and study of the geomorphosites. This will highlight the scientific and aesthetic value of the geomorphosites of the latter. The results of this study will complete the database of the Mount Bambouto geomorphosites, which will be useful to decision-makers for the implementation sustainable geotourism and geoconservation strategies in that locality.

Literature review

Scientific works on Mount Bambouto have focused more on the study of its dynamics and its geomorphological, petrographic, geochemical, and geochronological characteristics. A summary of these works is presented later. Mount Bambouto is a shield volcano truncated by two calderas (occidental and oriental), formed by an intra-volcanic mechanism (Tchoua, 1974). Unlike other authors, Zangmo Tefogoum et al. (2020a) showed that Mount Bambouto has a single caldera set up by explosive activities followed by a series of collapses as reported by Youmen (1994). For these authors, volcanic activity in Mount Bambouto is marked by three eruptive processes: effusive, responsible for flows; intrusive, with the formation of domes; and explosive, with the formation of ignimbrites. According to Tchoua (1974), explosive activities, characterised by ignimbritic phases, dominated the early stage of the history of Mount Bambouto volcanism. Later, this proposal was contradicted by Nono et al. (2003, 2004). In fact, for these authors, the presence of basaltic scoria and other basic xenoliths in the ignimbrites suggests that there was an initial phase of basaltic activity.

Kagou Dongmo et al. (2006, 2010) suggested that through the discovery of a strombolian pyroclastic cone at Totap-Taptsetsa on the north-eastern flank of Mount Bambouto, two distinct phases of strombolian activities exist in Mount Bambouto: a first ante-ignimbritic phase appearing as an enclave in the ignimbrites (Nono et al., 2003, 2004) and therefore older (around 21.12 Ma) and another, younger phase (0.5 Ma) at Totap-Tapsetsa, which would represent the last episodes of volcanism in Mount Bambouto. Between both phases, there was the formation of the caldera and the emplacement of rhyolitic and trachytic ignimbrites pointedly between 18.8 and 15.3 Ma, and the post-caldera formation of trachytic and phonolitic domes and ignimbritic deposits in the south of the massif, precisely between 15 Ma and 4.5 Ma. Ignimbrite deposits are from pyroclastic flows and constitute one of the harmful hazards around the globe. That is why Gountié Dedzo et al. (2020) determined the flow directions of ignimbrites using data from magnetic susceptibility anisotropy. This method enabled them to produce the pyroclastic flow hazards and risk maps. The latter show that the localities inside the calderas and those on the slopes of the volcano are arguably more exposed. Moreover, these authors stated that there are several domes on the Mount Bambouto caldera floor and slopes. According to Zangmo Tefogoum et al. (2019), they are classified as dome-flows and intrusive domes. The latter are relevant geomorphosites that can be promoted through the implementation of geotourism. Furthermore, the complexity of the Mount Bambouto caldera prompted Zangmo Tefogoum et al. (2020a) to classify it using the model of Geyer and Martí (2008). It emerged that this caldera was formed by a piecemeal collapse ruled by continental rifting of an extensional type (CR-EXT). According to these authors, the caldera is alkaline and ignimbritic. Since this caldera underpins post-caldera domes of a trachytic and phonolitic nature, it is of S-type (Vents scattered within the caldera) and MS-type (Multiple vents at the caldera margins).

Materials and methods

Several methods have been developed to study geomorphosites across the world. Some are only applicable in specific geographical contexts. For a significant number of years, the method of Reynard et al. (2016) has been applied.

According to these authors, the selection of geomorphosites in a specific environment must be done through the following important steps.

1. Define the main geomorphological contexts: the principal active and fossil geomorphological processes present in the area are listed. In our work, we applied a morphogenetic division related to groups of processes, namely volcanic, fluvial, and tectonic.
2. Based on a literature survey, consult cartographic material (topographic and geologic maps, digital terrain models, and aerial photographs), a field survey, other existing inventories and knowledge to establish a first list of landforms. The data, namely location, short description including geological features, accessibility, and educational interest and references were stored in a database and classified following the geomorphological contexts.
3. Classify the landforms following two sets of criteria: a spatial criterion (representative vs. rare landforms) and a temporal criterion (active vs. inherited landforms).
4. Select potential geomorphosites, which leads to a final list that is representative of the principal geomorphological contexts and various stages of the regional morphogenesis (inherited and active landforms).

In the lab we devoted our work to productions of the plates of figures and the mapping of the study area and its geomorphosites. A DEM has been downloaded on the OpenTopography website and imported into ArcGIS 10.4.1 where we used the arctoolbox parameter to develop the hillshade. Furthermore, we utilised Google Earth Pro software for the first inventory and the determination of some geographical data of geomorphosites. Moreover, some features have been digitised and imported to ArcGIS 10.4.1. In addition, the geographical coordinates of each geomorphosite were processed in Excel 2016 and later on they were imported as well as digitised features into ArcGIS 10.4.1 and superimposed on the hillshade to construct the geomorphosites map. Afterwards, the pictures taken from the field survey were compiled, sorted and exported in Adobe Illustrator CC 2020 to create various plates of figures. Finally, the results were interpretated and the paper was written.

Results and discussions

The main geomorphosites scattered on the Mount Bambouto slopes encompass domes, dykes, Totap cone (volcanic landforms), escarpments (tectonic landforms), waterfalls, potholes, and depression lakes (fluvial landforms). Figure 7.2 highlights the spatial distribution of these landforms.

Domes

Domes (around five) are the most widespread geomorphosites on the Mount Bambouto slopes. Two main types observed are dome-casts and extrusion domes (Youmen, 1994). According to Zangmo Tefogoum et al. (2019), their setting-up went through several steps that are summed up as follows: During the evolution stages of Mount Bambouto, several fissures were generated in the country rocks through the ascending magma pressure and magmas flowed through some of these fissures. The supply of magma to the fissure was progressive, until its total filling. The erected dome expanded with time through a continuous internal supply. Magmatic materials flowed down under its weight and the shield volcano slopes. Later, the

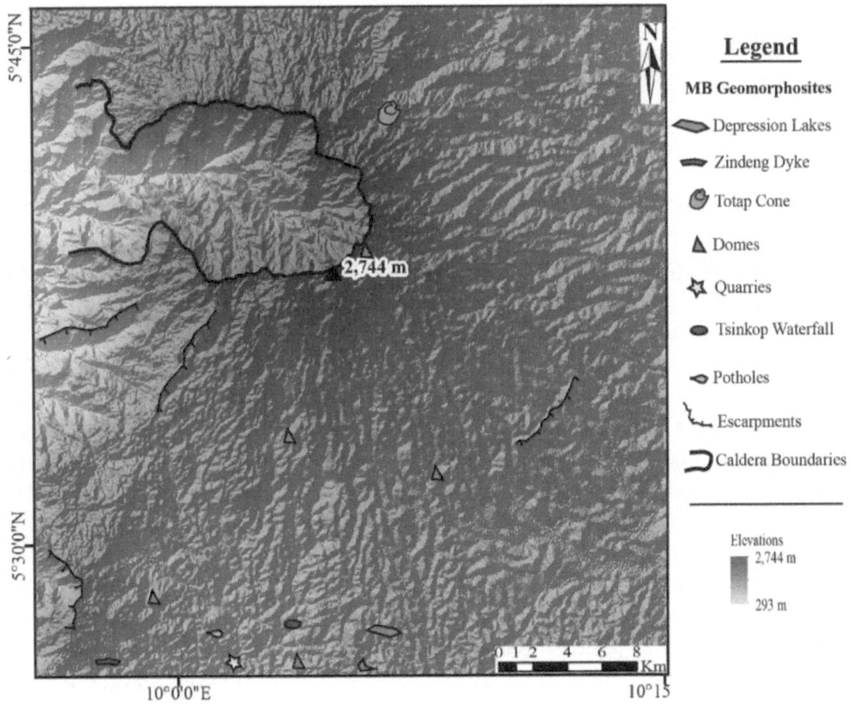

Figure 7.2 Spatial distribution of geomorphosites on Mount Bambouto.

Source: OpenTopography

dome cooled down and irregular prisms appeared on its surface. Most of these domes are conical, with a base that is wider than the rounded top. They also have grassland-like vegetation cover and a few rare shrubs.

Extrusive domes dominate the crest line and floor of the caldera, while the slopes are dominated by dome flows. The domes are fairly widely distributed around the volcano (Figure 7.3 a, b). Their considerable number reflects the degree of fracturing of Mount Bambouto and the significant number of magma flows that have governed its history. Heavy urbanisation, accompanied by a frantic race to urbanise several localities, has led to the opening of several rock quarries on the domes. The most significant is located at Foréké (south-west of the town of Dschang). It is a basalt dome covering an area of around 702 ha and rising to a height of around 1,475 m (Figure 7.3 c, d, e, and g). Nearly 55 per cent of the dome has already been breached through the opening of a quarry, with a face that is almost 235 m wide and almost 65 m high. This quarry is now considered an active mining geomorphosite. In fact, in quarries, everyone can look inside the Earth, see the internal structure of rock massifs and realise the connections that are not visible on the surface (Emil, 2020). This site highlights the internal structure of the volcanic formations, in particular the presence of varied colonnades at the base and false colonnades in the upper part, both separated by several horizontal joints (Figure 7.3e). The blocks detached from these gigantic prisms are polygonal, with five to six faces (Figure 7.3f). This site is a natural laboratory for training undergraduate and postgraduate students and an asset for the development of research.

Dykes

Dykes are more widespread in the northern parts of Cameroon than in the southern parts (Mey Atour et al., 2021; Gountié Dedzo et al., 2023). On the Mount Bambouto slopes, the benchmark site is the Zindeng dyke (Figure 7.4a). It is about a 300 m long, 100 m wide outcrop (Njanko et al., 2020), and 1 to 3 m high. It was emplaced during the late-Miocene Messinian age (6.61 ± 0.17 Ma– Nkouathio et al., 2008) on the Mid-Miocene basalts (12.52 ± 0.29 Ma) that cover the pan-African granitic basement. It is made up of porphyritic phonolite and it plays an important role in understanding the volcanic processes of Mount Bambouto, because it expresses the extrusive processes that are not common to all volcanic provinces around the world, which generally showcase explosive and effusive processes.

Escarpments

The domes, sometimes coalescing, coupled with the relatively more or less viscous flows, give a fairly heterogeneous appearance to the various slopes. In places, there are breaks in the slope that define large escarpments corresponding to flow fronts. The ignimbritic flow front (Figure 7.4b, 7.4c) can be seen on the road to Zindeng (south of Dschang).

On the south-western slopes of Mount Bambouto, there is an ignimbritic flow front that is several dozen metres wide and several metres high, defining

Figure 7.3 Pictures of geomorphosites and features on Mount Bambouto slopes: a & b) Domes; c, d, & e) Columnar joints; f) Polygonal block of basalt; g) Mudflow at the Foréké Dschang Quarry.

Figure 7.3 (Continued)

Figure 7.3 (Continued)

Figure 7.3 (Continued)

a major escarpment that forms one of the banks of the main watercourse running through the area. This geosite is criss-crossed by numerous diffuse fissures that eventually isolate falling or toppling boulders at its base. It is therefore a site where you can experience the dynamics of the landscape, particularly mass movements such as falling boulders and debris, which are surface phenomena. In some places, charred wood has been found enclosed in these ignimbrites. This type of enclave is also observed on the horizons of certain soils developed on ignimbrites (Figure 7.4d).

Totap cone

The Totap volcano (5°43'15"N, 10°07'50"E) is a SW–NE elongated volcano (2,100 × 1,700 m), covering 2.7 km^2, at the north-eastern upper slopes of Mount Bambouto. Its crater, 750 × 350 m large with a depth of 100 m, is opened to the north-east along the downhill slope. The age of this volcano is 0.480 ± 0.014 Ma. Thus, there was a 4 Ma quiescence of Mount Bambouto activities after the 4.5 Ma last basanitic eruption at the eastern flank. The upper flank basanite eruption of the Totap can be a fourth stage in the recent quaternary. On the road embankment, several fairly banded deposits can be clearly observed. This is the general structure found on the impressive number of stratovolcanoes around the world. There is the

Figure 7.4 Pictures of some geomorphosites on the Mount Bambouto slopes: a) Zindeng dyke; b & c) Escarpments; d) Enclave of wood in Zindeng ignimbrites; e, f, & g) Totap cone and features.

Figure 7.4 (Continued)

Figure 7.4 (Continued)

Figure 7.4 (Continued)

alternation of fine and coarse grains of well-defined scoria that imbricate some volcanic bombs (Figure 7.4e, f, and g). This structure can be considered testimony of the impressive number of volcanic explosions that ruled the formation of the Totap volcano at the north-east slopes of Mount Bambouto. It is the only strombolian cone in the whole Bambouto volcanic system. That is why it deserves more attention from a scientific point of view.

Waterfalls

Some slope failures offer opportunities for the formation of waterfalls. One example is the Tsinkop waterfall (in the centre of Dschang), where the water flows over a basalt plateau (Figure 5a, b, c). The waterfall is around 17 m high and falls on polygonal blocks detached from the basaltic flow face.

Depression lakes

The dynamics of Mount Bambouto have created small depressions where rainwater and run-off accumulate to form small natural lakes, notably Lake 8 and Lake Municipal. Lake 8 (Figure 5d, e) has been used for a long time as the main water supply source in the production of sand in the locality. As a result of the sand production, the lake water quality became bad and the aquatic fauna was destroyed. This was related to the progressive accumulation of clays in the lake. Thus, the fishing activity that in the past secured a livelihood to craftsmen is no longer pursued. Moreover, clay accumulation has over time reduced the lake's surface. In the past, the lake's surface area was around 15,650 m^2; nowadays it is about 6,080 m^2.

Lake Municipal is elliptical in shape and stretches in a west–east direction. It measures around 114,293 m^2 and, unlike Lake 8, is protected by the municipality. The overflow from this lake continuously feeds the Tsinkop waterfall. It also serves as a venue for weddings, recreational activities, and watersports, with a number of pirogues available for visitors to sail around and enjoy the lake. The museum built on its northern shore and the hotels found in its neighbourhood attract a large number of tourists.

Potholes

On the slopes of Mount Bambouto, particularly in the locality of Dschang, a section of the bed of one of the major rivers in the locality contains giant potholes. These geomorphosites are developed on ignimbrites and are quite diverse in terms of shape, diameter, and depth. Their shapes are circular, sub-circular, elliptical, and arbitrary. Sub-circular to circular shapes have diameters ranging from 30 cm to almost 1 m. Elliptical and arbitrary shapes vary in size from less than 1 m in the minor axis to several metres in the major axis. The depths are relatively shallow, with dimensions of less than 50 cm (Figure 7.5). These giant potholes are testimonies of the relevant dynamics of the local streams in Dschang city.

In the locality of Dschang, where the giants' pots were dug on ignimbrites, other forms were recorded in addition to the sub-circular and circular forms. This shows that lithology also plays an important role in the erosion of rocky rivers (Abbott & Pottratz, 1969; Wohl, 2008). Since ignimbrites are formations that are less resistant to alteration than surrounding gneisses and other volcanic and plutonic rocks, their alteration is likely to have been diffuse and to have given rise to several multiform holes.

Mount Bambouto slopes geomorphosites result from volcanic, tectonic, and fluvial processes. Therefore, they are systems that resulted from the interaction of inactive and active agents over time and territory (Comanescu et al., 2011). Many geomorphosites are characterised by a high cultural value owing to their connection with human activities and settlements (Panizza & Piacente, 2003; Reynard et al., 2007) as well as their geohistorical importance (Bollati et al., 2012). On the slopes of Mount Bambouto, these sites are undeniable components of the geodiversity of this part of the Volcanic Line of Cameroon. Through their ecosystem services, in particular regulating services, supporting services, provisioning services, and cultural services, Bamenda Mountains have several assets for human activities, such as the calderas of the Bamenda Mountains (Zangmo Tefogoum et al., 2022b). Several Mount Bambouto slopes that are dominated by domes are sites of intense agricultural and pastoral activities practised by local populations. Many of the domes

Figure 7.5 Pictures of some geomorphosites on the Mount Bambouto slopes: a, b & c) Tsinkop waterfall; d & e) Dschang Municipal Lake; f & g) Potholes.

Figure 7.5 (Continued)

Figure 7.5 (Continued)

Figure 7.5 (Continued)

provide the opportunity to open rock quarries for road and building infrastructure. However, since the 1990s, a growing interest for the heritage value of geology and geomorphology has been observed in several parts of the world in relation to geo-conservation (Gray, 2004; Burek & Prosser, 2008), geotourism (Dowling & Newsome, 2006), and geopark issues (Zouros, 2004). Certain geomorphosites on the slopes of Mount Bambouto – notably the Totap cone, the potholes, the columnar joints of the Foréké quarry and the Lac Municipal – are geological heritage sites endowed with assets for educational and geotourism activities. Pelfini and Bollati (2014) demonstrated that both active and passive landforms, and therefore geomorphosites, are of great interest for scientific, educational, and tourism purposes. These geomorphosites bear great importance in understanding the paleogeomorphological evolution of the local area (Comanescu et al., 2011). However, very few of these sites are recognised as such and exploited. Yet the promotion of geological and geomorphological heritage is growing in importance for educational initiatives (Pelfini & Bollati, 2014) worldwide. In order to express the dynamics of landscape systems, Foréké dome, Totap cone, Zindeng dyke, and potholes have been distinguished (Pelfini & Bollati, 2014).

Active geomorphosites allow for the visualisation of geomorphological processes in action (Reynard, 2004). Passive geomorphosites testify to past environments and processes; they are an archival record of the history of the Earth

(Reynard, 2005). Evolving passive geomorphosites are landforms created through processes that are no longer in action, and are reshaped by currently active processes (Pelfini & Bollati, 2014). Geosites constitute a sequence of temporary dynamic equilibriums of an evolutive system (Grecu, 2017). In the study area, active sites such as the Foréké quarry, the Tsinkop waterfall, and the Tchoualé giant potholes enable surface processes to be visualised and certain surface morphogenetic processes to be understood, respectively the human influence on the evolution of quarries, water erosion on escarpments, and the dynamics of watercourses and their impact on the evolution of the giant potholes. Their educational value is relatively high, which is why some of these sites are much in demand for student field trips. Even active quarries such as those at Foréké have these important values; many authors drawn attention to the values of abandoned quarries in their works (Stefano & Paolo, 2017; Baczynska et al., 2018)

Some geomorphosites on Mount Bambouto slopes are vulnerable and their disappearance can be attributed to the same process that produced them. Indeed, according to Grecu (2017), the dynamics and evolution of these geomorphosites lead to the deterioration and destruction of the geosite, to the generation of a new geosite, or to the disappearance of the geosite. The Foréké quarry is the result of the transformation of a basaltic dome. It allows us to observe the juxtaposition of normal and diffuse prisms, which are very important for understanding the cooling processes of basic magmatic material. Unfortunately, the unbridled exploitation of the rocks on this site could lead to its total destruction. Some sites, notably the escarpments and the Tsinkop waterfall, may also be destroyed by rock falls. Thus, geoconservation strategies must be established and implemented for those vulnerable geomorphosites.

In Cameroon, geotourism is not mentioned in the documents of the Ministry of Tourism and Leisure or in school curricula and research institutions. This is linked to the lack of awareness shown by decision-makers and the general public in Earth Sciences and hence in geological heritage (Zangmo Tefogoum et al., 2017). This work shows that the slopes of Mount Bambouto are dominated by several geomorphosites, most of which are volcanic in origin, providing information about the local geology and geomorphology. These are real assets for the implementation of geotourism. Geotourism is a means of combating poverty, but above all of preserving and protecting the unique natural wealth of Cameroon's geological landscape.

Conclusion and recommendations

The Mount Bambouto slopes are rich in geomorphosites. They are characterised by a number of diverse geomorphosites, mostly of volcanic origin. These include domes, dykes, the Totap cone, escarpments and waterfalls, potholes, and depression lakes. Some of these sites are passive, while others are active. Their cultural, geoeducational, and geotourism value is quite significant. That is why these geomorphosites must be promoted and preserved in order to foster recreation activities and long-term field training of students from national institutions. Mount

Bambouto geomorphosites and others in Cameroon deserve more attention. For this reason, the following recommendations are made:

- Experts in the domain through the support of the government must make a sustained inventory and the promotion of geological heritage sites in Cameroon.
- Stakeholders should fund infrastructures projects that could lead to the improvement of access conditions to certain geomorphosites.
- Stakeholders should reinforce protection of, and limit uncontrolled access to, Municipal Lake. In addition, the pollution of the lake's waters should be combated to protect the health of visitors.
- The stakeholders, in collaboration with experts in the domain of geoheritage, must develop and plant interpretive panels on most of the geomorphosites.

References

Abbott, A.T., Pottratz, S.W. (1969). Marine pothole erosion, Oahu, Hawaii. *Pacific Science*, 23(3), 276–290.

Baczynska, E., Lorenc, M.W., Kazmierczak, U. (2018). The landscape attractiveness of abandoned quarries. *Geoheritage*, 10(2), 271–285.

Bollati, I., Pelfini, M., Pellegrini, L. (2012). A geomorphosites selection method for educational purposes: A case study in Trebbia Valley (Emilia Romagna, Italy). *Geografia Fisicae Dinamica Quaternaria*, 35(1), 23–35.

Burek, C.V., Prosser, C.D. (Eds.). (2008). *The history of geoconservation. Special publication 300*. London, England: The Geological Society, p. 312.

Burke, K. (2001). Origin of the Cameroon Line of volcano-capped swells. *Journal of Geology*, 109, 349–362.

Bussard, J. (2014). Protection et valorisation du patrimoine géomorphologique du Parc naturel régional Gruyère Pays-d'Enhaut États des lieux et perspectives. Mémoire de Master, Université de Lausanne, Lausanne, Suisse, p. 120.

Cai, Y., Han, J., Wu, F., He, G. (2023). Promoting geotourism in Dunhuang UNESCO global Geopark. *Geoheritage*, 15, 22. https://doi.org/10.1007/s12371-023-00796-y.

Carton, A., Cavallin, A., Francavilla, F., Mantovani, F., Panizza, M., Pellegrini, G.G., Tellini, C. et al. (1994). Ricerche ambientali per l'individuazione e la valutazione dei beni geomorfologici – metodi ed esempi. *Il Quaternario*, 7(1), 365–372.

Comanescu, L., Nedelea, A., Dobre, R. (2011). Evaluation of geomorphosites in Vistea Valley (Fagaras Mountains-Carpathians, Romania). *International Journal of the Physical Sciences*, 6(5), 1161–1168. www.academicjournals.org/IJPS; https://doi.org/10.5897/IJPS10.384.

Déruelle, B. (1994). Le magmatisme de la "Ligne du Cameroun": Le bilan positif des deux dernières décennies de coopération franco-camerounaise. *Pangéa*, 20, 30–36.

Dincă, I. (2007). The theme of landscape, between the education of the tastes and themes of the achievement tourists-exercise on the inventory of the landscapes of the Department of Bihor (Romania)/Le thème du paysage, entre l'éducation des goûts et le profit thématique des touristes. Exercise sur l'inventaire des paysages du Département de Bihor (Roumanie). *Anuario Turismo y Sociedad*, 8, 83–105.

Dowling, R.K., & Newsome, D. (Eds.). (2006). *Geotourism. Sustainability, impacts and management*. Amsterdam, Netherlands: Elsevier.

Emil, S. (2020). Geotouristic potential of former quarries in northern bohemia. *6th World Multidisciplinary Earth Sciences Symposium, IOP Conference Series: Earth and*

Environmental Science, 609, 012079. IOP Publishing. https://doi.org/10.1088/1755-1315/609/1/012079.

Geyer, A., Martí, J. (2008). The new worldwide collapse caldera data base (CCDB): A tool for studying and understanding caldera processes. *Journal of Volcanology and Geothermal Research*, 175, 334–354. https://doi.org/10.1016/j.jvolgeores.2008.03.017.

Gountié Dedzo, M., Klamadji, M.N., Zangmo Tefogoum, G., Chako-Tchamabé, B., Asaah, A.N.E., Tchameni, R. (2023). Major, trace elements and Sr – Nd – Pb isotopes systematics of mafic dykes from the Figuil (northern Cameroon) and Léré (southwestern Chad) areas. *Geological Journal*, 59(3), 784–810. https://doi.org/10.1002/gj.4892.

Gountié Dedzo, M., Zangmo Tefogoum, G., Chako Tchamabé, B., Fozing, E.M., Njonfang, E., Kamgang, P. (2020). Mapping of pyroclastic density currents hazards and assessment of related risks by AMS technique in the West Cameroon highlands: Case of Bambouto and Bamenda volcanoes. *Journal of Geography, Environment and Earth Science International*, 24(2), 39–60.

Grandgirard, V. (1995). Méthode pour la réalisation d'un inventaire de géotopes géomorphologiques. *UKPIK, Cahiers de l'Institut de Géographie de l'Université de Fribourg*, 10, 121–137.

Grandgirard, V. (1997). Géomorphologie, protection de la nature et gestion du paysage. Thèse de Doctorat, Institut de Géographie, Université de Fribourg.

Gray, M. (2004). *Geodiversity: Valuing and conserving abiotic nature*. Chichester, England: Wiley & Sons.

Grecu, F. (2017). Interconditionality geomorphosites and natural hazards. *Riscuri Si Catastrofe*, 20(1), 41–51. https://doi.org/10.1016/j.jafrearsci.2019.103728.

Halliday, A.N., Dickin, A.P., Fallick, A.E., Fitton, J.G. (1988). Mantle dynamics: A Nd, Sr, Pb and O isotopic study of the Cameroon Line chain. *Journal of Petrology*, (1) 29, 181–211.

Hooke, J.M. (1994). Strategies for conserving and sustaining dynamic geomorphological sites. In Geological and *landscape conservation*. Halloran, D. O'., Green, C., Harley, M., & Knill, S.M.J. (Eds.), London, England: The Geological Society, pp. 191–195.

Kagou Dongmo, A., Nkouathio, D.G., Nono, A., Kwekam, M., Tematio, P., Tchoua, F.M., Pouclet, A., Boudier, J.L. (2006). Discovery of strombolian cone at Totap-Tapsetsa, north-east slope of Mount Bambouto: Implications in the petrogenesis and volcanic activity of Bambouto volcano (Cameroon Line). *Journal of African Earth Sciences*, 57(1–2), 96–108.

Kagou Dongmo, A., Nkouathio, D.G., Pouclet, A., Bardintzeff, J-M., Wandji, P., Nono, A., Guillou, H. (2010). The discovery of Late Quaternary basalt on Mount Bambouto: Implications for recent widespread volcanic activity in the southern Cameroon Line. *Journal of African Earth Sciences*, 57, 87–108.

Kenfack, P.L., Tematio, P., Kwekam, M., Ngueutchoua, G., Njike, P.R. (2011). Evidence of a Miocene volcano-sedimentary lithostratigraphic sequence at Ngwa (Dschang Region, West Cameroon): Preliminary analyses and geodynamic interpretation. *Journal of Petroleum Technology and Alternative Fuels*, 3(2), 25–34.

Krishnanand. (2019). Interface of geodiversity, geomorphosites, geotourism and seasonal economy in Lahaul and Spiti, Himachal Pradesh, India. *Landform Analysis*, 38, 25–34. https://doi.org/10.12657/landfana-038-003.

Lopes, O.A., Diawara, B. (2023). UNESCO and global geoparks. *Workshop on UNESCO Global Geoparks in Africa*, 21–25 September 2023, Windhoek, Namibia.

Marzoli, A., Piccirillo, E.M., Renne, P.R., Bellieni, G., Iacumin, M., Nyobe, J.B., Tongwa, A.T. (2000). The Cameroon volcanic line revisited: Petrogenesis of continental basaltic magma from lithospheric and asthenospheric mantle source. *Journal of Petrology*, 41(1), 87–109.

Mey Atour, M., Bardintzeff, J.M., Fagny Mefire, A., Nkouandou, O.F., Tchameni, R., Dinamou, A. (2021). Petrology and geochemical framework of dolerites dykes of Temte´,

North Cameroon, Central Africa. *Acta Geochim*, 40(3), 287–299. https://doi.org/10.1007/s11631-020-00432-4.

Nana, R. (2001). Pétrologie des Péridotites en enclaves dans les basaltes alcalins récents de Nyos: Apport à la connaissance du manteau supérieur de la Ligne du Cameroun. Thèse d'Etat, Université Yaoundé 1, Cameroun, p. 284.

Newsome, D., Dowling, R.K. (2010). *Geotourism: The tourism of geology and landscape*. Oxford, England: Good Fellow Publishers Ltd, p. 260.

Ngako, V., Njonfang, E., Aka, F.T., Affaton, P., Metuk Nnange, J. (2006). The North-South Paleozoic to Quaternary trend of alkaline magmatism from Niger – to Cameroon: Complex interaction between hotspots and Precambrian faults. *Journal of African Earth Sciences*, 45, 241–256.

Njanko, T., Gountié Dedzo, M., Tamen, J., Bella Nke, E.B., Kadji Kouémo, O.S., Piankeu Doumsab, G., Tchakounté, J., Fozing, E.M. (2020). Emplacement of the Zindeng phonolitic lava flow (West-Cameroon) in the Cameroon volcanic line: Constraints from the anisotropy of magnetic susceptibility (AMS). *Journal of African Earth Sciences*, 162, 103728.

Njonfang, E., Nono, A., Kamgang, P., Ngako, V., Tchoua, F.M. (2011). Cameroon line alkaline magmatism (Central Africa): A reappraisal. *Geological Society America Special Paper*, 478, 173–191.

Nkouathio, D.G., Kagou Dongmo, A., Bardintzeff, J.-M., Wandji, P., Bellon, H., Pouclet, A. (2008). Evolution of volcanism in graben and horst structures along the Cenozoic Cameroon Line (Africa): Implications for tectonic evolution and mantle source composition. *Mineralogy and Petrology*, 94, 287–303.

Nono, A., Njonfang, E., Kagou Dongmo, A., Nkouathio, D.G., Tchoua, F.M. (2004). Pyroclastic deposits of the Bambouto volcano (Cameroon Line, Central Africa): Evidence of an initial strombolian phase. *Journal of African Earth Sciences*, 39, 409–414.

Nono, A., Nkouathio, D.G., Gountié Dedzo, M., Njonfang, E., Kagou Dongmo, A., Tchoua, F.M. (2003). Zonal and vertical variations in welding rate and composition of ignimbrites in the Bambouto volcano (Cameroon Line, Central Africa): Volcanological importance. *Geophysical Research Abstracts*, 5, 14541.

Panizza, M. (2001). Geomorphosites: Concepts, methods and example of geomorphological survey. *Chinese Science Bulletin. Supplementband*, 46, 4–6.

Panizza, M., Piacente, S. (1993). Geomorphological assets evaluation. *Zeitschrift für Geomorphologie. Supplementband*, 87, 13–18.

Panizza, M., Piacente, S. (2003). *Geomorfologia culturale*. Bologna: Pitagora Editrice, p. 350.

Pelfini, M., Bollati, I. (2014). Landforms and geomorphosites ongoing changes: Concepts and implications for geoheritage promotion. *Quaestiones Geographicae*, 33(1), 131–143. https://doi.org/10.2478/quageo-2014-0009; ISSN 0137-477X.

Reynard, E. (2004). Géotopes, géo(morpho)sites et paysages géomorphologiques. In *Paysages géomorphologiques. Institut de Géographie*, Vol. 27. Reynard, E. & Pralong, J.-P. (Eds.). Lausanne: Travaux et Recherches, pp. 123–136.

Reynard, E. (2005). Géomorphosites et paysages. *Géomorphologie Relief, Processus, Environnement*, 3, 181–188.

Reynard, E., Fontana, G., Kozlik, L., Scapozza, C. (2007). A method for assessing scientific and additional values of geomorphosites. *Geographica Helvetica*, 62(3), 148–158.

Reynard, E., Perret, A., Bussard, J., Grangier, L., Martin, S. (2016). Integrated approach for the inventory and management of geomorphological heritage at the regional scale. *Geoheritage*, 8(1), 43–60. https://doi.org/10.1007/s12371-015-0153-0.

Rivas, V., Rix, K., Frances, E., Cendrero, A., Brunsden, D. (1997). Geomorphological indicators for environmental impact assessment: consumable and non-consumable geomorphological resources. *Geomorphology*, 18(3–4), 169–182.

Shayan, S., Sharifikia, M., Zare, G.R. (2011). Evaluating the geomorphotourism potential of landforms based on Praloong method (Case study: Darab Township). *Arid Regions Geographic Studies*, 1(2), 73–91.

Stefano, M., Paolo, S. (2017). Abandoned quarries and geotourism: An opportunity for the Salento Quarry District (Apulia, Southern Italy). *Geoheritage*, 9(4), 463–477.

Suh, C.E., Sparks, R.S.J., Fitton, J.G., Ayonghe, S.N., Annen, C., Nana, R., Luckman, A. (2003). The 1999 and 2000 eruptions of Mount Cameroon: Eruption behaviour and petrochemistry of Lava. *Bulletin of Volcanology*, 65(4), 267–281. https://doi.org/10.1007/s00445-002-0257-7.

Tabod, C.T., Fairhead, J.D., Stuart, G.W., Ateba, B., Ntepe, N. (1992). Seismicity of the Cameroon Volcanic Line, 1982–1990. *Tectonophysics*, 212, 303–320.

Tamen, J., Nkoumbou, C., Mouafo, L., Reusser, E., Tchoua, F.M. (2007). Petrology and geochemistry of monogenetic volcanoes of the Barombi Koto volcanic field (Kumba graben, Cameroon Volcanic Line): Implications for mantle source characteristics. *Comptes Rendus de l'Académie des Sciences*, 339(13), 799–809.

Tchoua, F.M. (1972). Sur la formation des calderas des monts Bambouto (Cameroun). *Comptes Rendus de l'Académie des Sciences*, 274, 799–801.

Tchoua, F.M. (1974). Contribution à l'étude géologique et pétrographique de quelques volcans de la "Ligne du Cameroun" (Monts Manengouba et Bambouto). Thèse d'Etat, Université Clermont-Ferrand, France, 337 pp + Biblio + 75 figures.

Tematio, P., Kengni, L., Bitom, D., Hodson, M., Fopoussi, J.C., Leumbe, O., Mpakam, H.G., Tsozué, D. (2004). Soils and their distribution on Bambouto volcanic mountain, West Cameroon Highland, Central Africa. *Journal of African Earth Sciences*, 39, 447–457.

Tematio, P., Tsafack, E.I., Kengni, L. (2011). Effects of tillage, fallow and burning on selected properties and fertility status of Andosols in the Mounts Bambouto, West Cameroon. *Journal of Agricultural Science*, 3, 334–340.

Wantim, M.N., Kervyn, M., Ernst, G.G.J., Del Marmol, M.-A., Suh, C.E., Jacobs, P. (2013). Morpho-structure of the 1982 Lava flow field at Mount Cameroon Volcano, West-Central Africa. *International Journal of Geosciences*, 4, 564–583.

Wohl, E.E. (2008). The effect of bedrock jointing on the formation of straths in the Cache la Poudre River drainage, Colorado Front Range. *Journal of Geophysical Research*, 113, F01007.

Youmen, D. (1994). Evolution volcanologique, pétrologique et temporelle de la caldeira des monts Bambouto (Camerou). Thèse, Université Christian Abrecht Kiel, Allemagne, 274 pp + 2 cartes.

Youmen, D., Schminke, H.U., Lissom, J., Etame, J. (2005). Données géochronologiques: Mise en évidence des différentes phases volcaniques au Miocène dans les monts Bambouto (Ligne du Cameroun). *Sciences, Technologie et Développement*, 11(1), 49–57.

Zangmo Tefogoum, G., Dongmo Kagou, A., Nkouathio, D.G., Dedzo Gountié, M., Kamgang, P. (2017). The volcanic geoheritage of the Mount Bamenda Calderas (Cameroon Line): Assessment for geotouristic and geoeducational purposes. *Geoheritage*, 9(3), 255–278. https://doi.org/10.1007/s12371-016-0177-0.

Zangmo Tefogoum, G., Kagou Dongmo, A., Nkouathio, D.G., Wandji, P., Gountié Dedzo, M. (2014). Geomorphological features of the Manengouba volcano (Cameroon line): Assets for potential geopark development. *Geoheritage*, 6(3), 225–239. https://doi.org/10.1007/s12371-014-0109-9.

Zangmo Tefogoum, G., Nkouathio, D.G., Kagou Dongmo, A., Gountié Dedzo, M. (2019). Typology of geotouristic assets along the south continental branch of the Cameroon volcanic line: Case of the Mount Bambouto's caldera. *International Journal of Geoheritage Parks*, 7, 111–128. https://doi.org/10.1016/j.ijgeop.2019.06.003.

Zangmo Tefogoum, G., Nkouathio, D.G., Kagou Dongmo, A., Gountié Dedzo, M. (2020a). The Caldera of Mount Bambouto: Volcanological characterization and classification. In *Volcanoes – updates in volcanology*. Nemeth, K. (Ed.). http://dx.doi.org/10.5772/intechopen.93694.

Zangmo Tefogoum, G., Quesada Román, A., Pérez Umaña, D. (2020b). Geomorphosites inventory in the Eboga Volcano, Cameroon: Contribution for geotourism promotion. *Géomorphologie: Relief, Processus, Environnement*, 26, 19–33. https://doi.org/10.4000/geomorphologie.14006.

Zangmo Tefogoum, G., Nouhou Dama, H.Z., Gountié Dedzo, M., Mafo Dongmo, I. (2022a). The granitic inselbergs of Lara (Panafrican Chain, Far-North Cameroon): Assets for the implementation of geotourism in Mountainous region. *Journal of Mountain Science*, 19(11), 3192–3212. https://doi.org/10.1007/s11629-021-6862-8.

Zangmo Tefogoum, G., Kagou Dongmo, A., Nkouathio, D.G., Gountié Dedzo, M. (2022b). The geodiversity of Lefo and Santa-Mbu Calderas (Bamenda Mountains, Cameroon Volcanic Line): Factor for socioeconomic activities. *International Journal of Geoheritage and Parks*, 7, 111–128. https://doi.org/10.1016/j.ijgeop.2022.08.009.

Zgłobicki, W., Gawrysiak, L., Kołodyńska-Gawrysiak, R. (2015). Gully erosion as a natural hazard: The educational role of geotourism. *Natural Hazards*, 79(1), 159–181. https://doi.org/10.1007/s11069-014-1505-9.

Ziem à Bidias, L.A., Ilouga, I.D.C., Moundi, A., Nsangou, A. (2020). Inventory and assessment of the Mbepit Massif Geomorphosites (Cameroon volcanic line): Assets for the development of local geotourism. *Geoheritage*, 12, 49. https://doi.org/10.1007/s12371-020-00471-6.

Ziem à Bidias, L.A., Njeuatchoua Kameni, L.H., Moundi, A., Kamgang, P. (2023). Geoheritage of the volcanic landscapes of FoumbotKouomboum region, Noun Plain, Cameroon: Geomorphological features and assessment of geomorphosites. *International Journal of Geoheritage and Parks*, 11, 464–482. https://doi.org/10.1016/j.ijgeop.2023.08.002.

Zouros, N. (2004). The European geoparks network. Geological heritage protection and local development. *Episodes*, 27, 165–171.

8 Enhancing geoscience training in Ghana's tertiary institutions through geoconservation strategies

Marian Selorm Sapah, Daniel Kwadwo Asiedu, Yvonne Sena Akosua Loh, and Bruce Kofi Banoeng-Yakubo

Introduction

Geological field studies have long been established as an integral component in the training of geoscientists (Butler, 2008; Whitmeyer & Mogk, 2009). Geoscience training typically refers to specialised, hands-on learning experiences aimed at developing specific skills and competencies for practical application in the field or in industry (Mogk & Goodwin, 2012). Geoscience training programs may include fieldwork, laboratory work, workshops, or internships, designed to teach students how to use geoscientific instruments, software, and methodologies effectively. The emphasis in geoscience training is often on practical problem-solving, data analysis, interpretation, and the application of geoscience knowledge to real-world scenarios (Kastens & Manduca, 2012a; LaDue & Pacheco, 2013). It can be viewed as the practical component of geoscience education which provides a broad theoretical foundation for a comprehensive understanding of the earth sciences through formal academic study. Outcrops and other rock exposures in the field provide context critical to the study of geological features such as geological structures and petrographic composition of rocks and understanding geological concepts and processes such as metamorphism, rock deformation, rock erosion, and sediment deposition (Kastens & Manduca, 2012b). It is also worth noting that most of the academic research and literature on geoscience education and training with the emphasis on geological field studies are located in developed countries such as the USA, Australia, Canada, UK, Germany, Netherlands, and China (Baoguo et al., 2016; Brown et al., 2018; King et al., 2021). These countries have robust academic communities dedicated to geoscience education and training and strong traditions of geological research and education, which has resulted in multifaceted outcomes, such as skilled workforce development, scientific advancement, global collaborations, environmental stewardship, adequate natural resource management, and other societal impacts (Capello & Howes, 2022; Sorkhabi, 2023). This has also led to the development and establishment of the concepts of geoheritage, geoconservation, and geotourism in these developed countries, which are integral components of sustainable land management and environmental protection, ensuring

DOI: 10.4324/9781032706764-8

the long-term preservation and enjoyment of geological treasures. This is unlike developing countries, particularly in Africa, where geoscience education and training is underdeveloped and not very well entrenched in both the formal and informal educational sectors, especially at basic levels (Martínez-Frías & Mogessie, 2012; North et al., 2020; Sapah et al., 2024). However, efforts have been made by various African countries to develop and strengthen geoscience education and training programs, as evidenced by peer-reviewed journal articles in which the status, challenges, and prospects of geoscience education and training in these regions are discussed (e.g. Arukwe et al., 2022; Sapah et al., 2024). To bridge this gap and accelerate the development of geoscience education and training in Africa, a multi-faceted approach, that addresses the various challenges and leverages opportunities for growth and advancement, is required. By addressing these factors and implementing strategic initiatives such as investment in infrastructure and resources, curriculum development and accreditation, capacity building, collaboration, and promotion of geoscience awareness and outreach, geoheritage and geoconservation, African countries can develop sustainable and high-quality geoscience education and training programs that contribute to sustainable development, environmental stewardship, and societal well-being.

Ghana is a country located in West Africa (Figure 8.1), with a population of 30.8 million as of 2021. The geology of Ghana can be summarised to consist of primarily very ancient crystalline basement rock, volcanic belts, and sedimentary basins, affected by periods of igneous activity and orogenic events (Kesse, 1985). The geology of Ghana is particularly interesting and important because of the age, tectonics, and lithologic characteristics of the supracrustal rocks. Apart from its rich geological history, there is also major economic interest in the geology of Ghana due to its mineral, oil, and water resources. In spite of the major advantages of the geology of Ghana, geoeducation is not well embedded in the formal and informal education system in Ghana (Sapah et al., 2024). The basic (primary and junior high) and senior high school curricula include very limited, or no, geoscience topics compared to other science disciplines such as biology, physics, and chemistry.

Geoscience subjects are only taught in a few tertiary institutions in Ghana, including the University of Ghana (UG), Kwame Nkrumah University of Science and Technology (KNUST), University of Mines and Technology (UMaT), C.K. Tedam University of Technology and Applied Sciences (CKT-UTAS), University for Development Studies (UDS), University of Energy and Natural Resources (UENR), University of Environment and Sustainable Development (UESD), and Radford University. They run geoscience programs with subjects such as geology, geomorphology, geophysics, geochemistry, engineering geology, mineral exploration, petroleum geoscience, palaeontology, hydrogeology, and environmental science, all of whom have strong field components. The geological field components of these programs are typically of two types: (1) class field trips of five to 10 days to locations with interesting geology – for most students this would be the crystallisation of the theories they have learnt in the classroom into the practical and (2) group fieldwork in an area of geological interest, where students typically produce

Figure 8.1 Map of Africa showing Ghana.

Source: Sapah et al., 2024

a geological map and understand the geology of the area. Many of these geological sites of interest visited by students are, however, vulnerable to urbanisation, infrastructure development, agriculture, over-use, and erosion. It is therefore essential to conserve these geological sites if we are to maintain these sites for research, education, and training. Geoconservation is not well established in Ghana, unlike in other parts of the world such as in the UK where it has been part of statutory nature conservation for more than 60 years (Prosser et al., 2013).

To initiate a conversation leading to action on geoconservation, particularly for geoscience education and training, geoconservation strategies following established best practices that can be applied in Ghana are discussed and recommended, with an emphasis on stakeholder institutions that should be involved and their tasks and contributions. This paper also outlines some geological sites of interest, as recommendations to the Ghana Geological Survey Authority (GGSA), for geoconservation by law, leading to their preservation as natural classrooms for geoeducation and geoscience training in Ghana. The following sections focus on the global understanding of geoconservation and geoscience training, geoconservation in Africa with the emphasis on Ghana, the status of geoscience training in tertiary institutions in Ghana with the Department of Earth Science, University of Ghana (UG) as an example, recommendations for methods of in situ geoconservation for geoscience training in Ghana, a discussion, and conclusion. The resources used in this research were scientific publications with subjects addressing geodiversity, geoheritage, geoconservation, and geoscience education and training globally, regionally, and locally.

The status of geoscience training in tertiary institutions in Ghana

Sapah et al. (2024) have reported that, out of 179 universities and colleges of education in Ghana, only 7 (i.e. 6 public and 1 private) offer geoscience-related programs. All 6 public universities have a geoscience program structure that includes high percentages of core geology courses (14–49 per cent) and applied geology courses (18–44 per cent). Of the 6 universities, 3 (UG, KNUST and UMaT) have geological field courses in their programs: short inter-semester field courses, and long vacation field mapping courses. Only the UG offers both types of field courses. KNUST and UMaT only offer short inter-semester field courses. For a detailed description of the status of geoscience training in tertiary institutions in Ghana, the geoscience program of the Department of Earth Science, UG, is used as a case study, as this is within the experience and expertise of the authors who help to develop and facilitate the program as well as supervise students.

The Department of Earth Science, UG, offers a four-year undergraduate BSc degree in earth sciences, with options in geology, applied geology and applied geophysics. Field-based courses form an important and integral part of the undergraduate curriculum. The University runs a semester system of education. There are two types of field courses: inter-semester field courses where students spend one to two weeks in a year in the field during the inter-semester break, under faculty supervision, to undertake field exercises that supplement their classroom studies (Figures 8.2 and 8.3) and long vacation field courses where third-year students who opt for a geology major in the final year undertake 4 to 6 weeks of independent field mapping studies during the long vacation (end of the second semester). In addition to these two types of field courses, there are several return-day trips to the field in order to reinforce geological concepts taught in class. These trips present a useful and interesting way to learn about the environment and geological processes. Other field trips include field excursions to large and small industrial firms throughout the entire country, or tunnel projects that are underway, as well as natural areas that require special measures regarding the environment. Final-year students also spend some time in the field to collect data for their independent dissertation work. Students do industrial attachment during the long vacation, to a governmental or private sector geoscience or related institution/company. Credit is

Figure 8.2 Undergraduate students of the Department of Earth Science, University of Ghana, undertaking intersemester field trips: a) Students measuring attitudes of an outcrop, b) Students observing and describing an outcrop, and c) Students listening to a field lecture.

contingent on the submission of a final report by students and an assessment report by industry. This internship course offers an opportunity for students with little or no experience to enter a professional working environment and work hands-on in their chosen field.

Figure 8.2 (Continued)

Figure 8.3 Images of sites from the Sekondian Group of rocks that need conservation, located in Sekondi-Takoradi: a) Ajua Shale Formation showing an example of a nonconformity, with sedimentary rocks overlying crystalline basement rocks. This stratigraphic feature is located in a fishing community where numerous human activities occur; b) Takoradi Shale Formation showing examples of symmetric ripple marks. This sedimentary structure is being affected by natural processes and human activities; c) (Photo taken in 2005) Image of petrified wood in the Sekondi sandstone; d) A recent photo (taken in 2023) of the petrified wood in c) displaced due to human activities in spite of efforts over the years to protect it.

Figure 8.3 (Continued)

Figure 8.3 (Continued)

Details of the inter-semester field exercises are as follows: The second-year students do about five to 10 days 'live-in' field exercises in a Phanerozoic sedimentary terrain, and they are given hands-on instructions on the recognition, identification, description, and interpretation of rocks and geological features. The third-year students undertake about one week 'live-in' field studies, with supervision, in an igneous/metamorphic terrain. Final-year students undertake one week-long field trip to hydrogeological, engineering, geoenvironmental, and mine sites. The trip introduces practical skills appropriate to the study of earth and environmental sciences. The course concentrates on interactions and feedback in the environment, including studies of geology, landforms, soil types, and water quality.

The field studies provide numerous benefits to the students. Just by being in the field, students have the opportunity to observe and investigate real geological concepts and processes, which is better than relying solely on textbook examples. In addition, field trips help students develop important skills such as critical thinking, problem-solving skills and lateral thinking, observation, data collection and data analysis, and working in teams. However, geological field studies are subjected to many challenges in a developing country such as Ghana, the most apparent being funding for field trips and the destruction of geological features due to urbanisation, and infrastructural developments.

Recommendations for methods of in situ geoconservation for geoscience training in Ghana

Geoheritage and geoconservation are not well-established practices in Ghana. As such, there are no known geological sites that have been conserved or are undergoing geoconservation for geoheritage, geotourism, or geoeducation. There is therefore a need for stakeholders in the geoscience community, such as academic institutions, the GGSA, the Ghana Institute of Geoscientists (GhIG), the Minerals Commission, the Environmental Protection Agency (EPA), government agencies such as the Ministry of Education (MoE), the Ministry of Lands and Natural Resources (MoLNR), the UNESCO Ghana Heritage Commission (GHC), and the Ghana Museum and Monuments Board (GMMB), as well as policy makers, to collaborate on developing and promoting geoheritage and geoconservation endeavours in Ghana, particularly for geological sites that are being used for geoscience training.

Globally established methodologies and best practices in geoheritage and geoconservation (Sharples, 1993, 2002; Brilha, 2016, 2018; Reynard & Brilha, 2017; García-Cortés et al., 2019; Crofts et al., 2020; IUGS-ICG, 2023) are a resource that can be applied in developing and promoting geoheritage and geoconservation in Ghana. These methods and established best practices as deduced from the literature mainly encompass (1) inventory and documentation: the systematic recording of geological sites, features, and landscapes and creation of comprehensive databases and catalogues, (2) assessment and evaluation: assessing the scientific, educational, cultural, and aesthetic value of geological sites, using criteria such as geological significance, rarity, representativeness, integrity, and vulnerability, (3) designation and recognition: official designation of sites as protected areas (e.g. geoparks, national parks, world heritage sites), (4) legal protection: enacting laws and regulations to protect geological sites from damage or destruction as well as establishing protected areas and defining buffer zones, (5) management: developing and implementing site-specific management plans, addressing issues such as visitor access, site maintenance, and risk management, (6) active conservation: physical conservation measures such as erosion control, vegetation management, and restoration of damaged sites, and also, monitoring and maintaining geological features to ensure long-term preservation, and (7) public education, outreach and promotion: developing educational programs and materials for schools, universities, and the general public, organising workshops, guided tours, and exhibitions, creating visitor centres, museums, and interpretative trails and using signage, brochures, and digital resources to explain geological features and their significance.

For the objective of this chapter, which is focused on the conservation of geological sites for geoscience training by tertiary institutions in Ghana, the following methods and practices are recommended. Table 8.1 summarises the different methods and practices, how they should be implemented, and who should implement them.

Table 8.1 Recommendations for methods of in-situ geoconservation for geoscience training in Ghana

Method/Practice	*Tasks*	*Activities*	*Stakeholders Responsible/Working group*
Inventory and Documentation	1. Identification of sites 2. Description of sites 3. Delineation of Sites	1. Literature review, 2. Interviews with experts 3. Field work	Academic Institutions, GGSA, Local Communities
Assessment and Evaluation	1. Assessment of Scientific value 2. Assessment of Educational values 3. Assessment of Risk of degradation	1. Literature review 2. Interviews with experts 3. Field work	Academic Institutions, GGSA, Local Communities
Designation and Recognition	1. Official designation of sites as protected areas (e.g., National Parks, National Geoparks, IUGS, UNESCO Geoparks). 2. Recognizing sites through awards or inclusion in national and international registers (e.g., Sites of Special Scientific Interest (SSSI)).	Preparation and submission of required documentation (i.e. an application dossier for e.g., GSSA, GHC parks and reservations project, GFC National parks and reservations. IUGS, UNESCO Geoparks projects)	GGSA, Academic institutions, GHC GFC, GhIG, MoE, MoLNR, EPA
Legal Protection	1. Enacting laws and regulations to protect geological sites from damage or destruction. 2. Establishing protected areas and defining buffer zones.	1. Enacting the GGSA ACT, 2016 ACT 298, Clause 26 (e) to protect geological sites from damage or destruction. 2. Other stakeholder institutions can provide legal designations, policies and regulations that can be used to protect these sites.	GSSA, District/ Municipal and Metropolitan Assemblies, Local Communities, EPAGHC, GFC

(*Continued*)

Table 8.1 (Continued)

Method/Practice	*Tasks*	*Activities*	*Stakeholders Responsible/Working group*
Management	1. Developing and implementing site-specific management plans. 2. Addressing issues such as visitor access, site maintenance, and risk management.	1. Preparing and implementing site specific management plans 2. Establishing protocols for accessibility, maintenance and risk management 3. Regular monitoring and maintaining of geological features through physical conservation measures such as erosion control, vegetation management, and restoration of damaged sites, to ensure long-term preservation.	Academic institutions, GGSA, GHC, GhIG, EPA, GFC, District/Municipal and Metropolitan Assemblies, Local Communities, MoE, MoLNR
Public Education, Outreach and Promotion	1. Developing educational activities and materials for schools, universities, and the general public. 2. Creating interpretative trails, using signage, brochures, and digital resources to explain geological features and their significance.	1. Developing, preparing and organizing educational activities and materials for schools, universities, and the general public 2. Creating and installing interpretative trails, using signage as well as preparing brochures and digital resources to explain geological features and their significance.	Academic institutions, GGSA, GHC, GhIG, EPA, MoE, GMMB District/ Municipal and Metropolitan Assemblies, Local Communities

Note: Ghana Geological Survey Authority (GGSA), the Ghana Institute of Geoscientists (GhIG), the Minerals Commission, the Environmental Protection Agency (EPA), government agencies such as the Ministry of Education (MoE), the Ministry of Lands and Natural Resources (MoLNR), the UNESCO's Ghana Heritage Commission (GHC), Ghana Forestry Commission (GFC), Ghana Museum and Monuments Board (GMMB), International Union of Geological Sciences (IUGS), and the United Nations Educational, Scientific and Cultural Organization (UNESCO)

Inventory and documentation

Conducting an inventory of geological sites for geoheritage and geoconservation is important for several reasons (García-Cortés et al., 2019). It provides a thorough understanding of the geological diversity and significance of an area. This knowledge is important for effective conservation planning and management. Inventories also serve as baseline data that can be used for monitoring changes and assessing the impact of natural and human activities on sites over time. An inventory helps prioritize sites based on their scientific, educational, cultural, and aesthetic values, ensuring that the most significant sites receive appropriate protection and management. Detailed documentation can be used to educate the public and raise awareness about the importance of geological heritage, promoting its appreciation, and support for conservation efforts. Furthermore, well-documented sites are more likely to receive legal protection and attract funding for their conservation, as detailed records demonstrate their significance and the necessity for their preservation. Inventories also serve as a valuable resource for researchers, facilitating the exchange of knowledge and best practices in the field of geoheritage and geoconservation.

To be able to carry out geoconservation for geoscience training in Ghana, there is a need to properly identify and document geological sites appropriate for this purpose in Ghana. This means geological sites with geological interests of scientific and educational value. This inventory should include the identification and detailed description, delineation, assessment, and legal protection of such sites. Data collection should be done through fieldwork by experts such as geologists, and spearheaded by the academic institutions that offer geoscience programs in Ghana, supported by the GSSA. This can be done on a region-by-region scale, using the administrative regional boundaries of the country. Geological sites that are already being used for geoscience training should be prioritised. Preceding such field missions, information on community involvement (for local knowledge of the area and land/property ownership issues), accessibility of the area (climate, weather, roads etc.), demography of the area (population, economy etc.) and infrastructural logistics (accommodation, restaurants etc.) must be available for efficient planning (Tavares et al., 2015; Adade Williams et al., 2020).

Identification and description of geological sites

Geological sites can be identified through literature review, interviews with experts, and field missions (Sharples, 1993; Brilha, 2016; Reynard & Brilha, 2017). A potential geological site for geoscience training should be an area with geological elements recognised mainly for their scientific and educational value. It should offer a diverse range of geological features that are well-documented and accessible, facilitating both research and hands-on learning experiences for students and researchers alike (Elkins & Elkins, 2007; Butler, 2008; Stokes et al., 2009).

The geological sites of interest currently being used for geoscience training by tertiary institutions in Ghana were areas of geological interest selected based on research on the geology of Ghana. The geology of Ghana is divided into five main provinces (Kesse, 1985), namely (1) the Paleoproterozoic Birimian basement and its associated Tarkwaian sediments occurring in the western portion of the country, (2) the Pan African (Neoproterozoic) Dahomiyide belt to the east, (3) the Neoproterozoic Voltaian in the central portions, (4) the Paleozoic-Mesozoic sedimentary cover in the coastal south, and (5) the Tertiary to recent sedimentary deposits scattered across the country. Other factors considered for the selection of these sites include lithology, stratigraphic sequence, geologic structures, sedimentary structures, fossils, and mineral and hydrocarbon potential. Figure 8.4 shows the simplified geological map of Ghana with the areas where geological sites of interest currently being used for geoscience training by the Department of Earth Science, UG, are located. These sites have not been described, delineated, or assessed for inventory towards geoconservation.

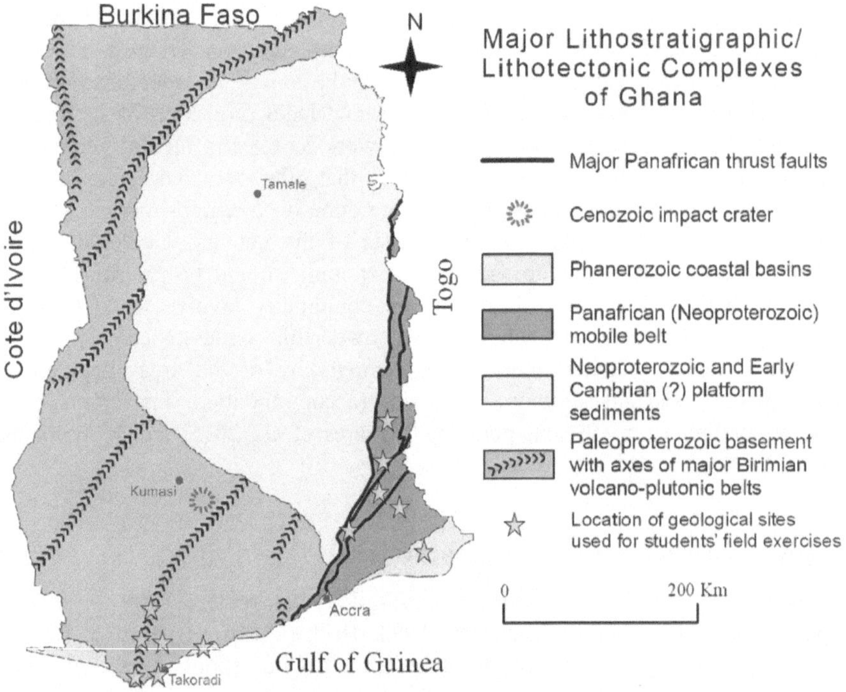

Figure 8.4 A simplified geological map of Ghana showing the location of sites visited during students' intersemester field training.

Source: Adapted from Agyei Duodo et al. (2009)

The description of these sites and new potential sites for inventory towards geoconservation should encompass:

- Location, including geographical coordinates
- Geological setting and domain
- Geological phenomena and/or processes
- Age of phenomenon or process
- Age of terrain or unit
- Geological features and structures
- Main geological interest of scientific and educational value
- Rarity
- The relationship of the site with other forms of natural diversity surrounding it

Figure 8.5 shows some geological sites of interest with the potential for geoscience training that should be considered for geoconservation.

Figure 8.5 Some interesting geological sites in Ghana that we think have scientific and educational value and should be conserved for geoscience training: a) Edikan Gold Mine, located in the Ashanti Gold Belt. It has interests for metallogeny, mineralogy, petrology, geochemistry, stratigraphy, geomorphology, tectonics, etc.; b) Lake Bosumtwi resulting from an asteroid impact 1.07 Ma ago. It is located within the Birimian Supergroup of Ghana. It has interests for astronomy, stratigraphy, geomorphology, paleontology, tectonics, petrology, geochemistry, metallogeny, mineralogy, hydrogeology, etc.; c) Wulin Mushroom rocks, found in Jirapa, Upper West Region, Ghana, have interests particularly for geomorphology; and d) Umbrella rock, found in Koforidua, Eastern Region, Ghana, has interests particularly for geomorphology.

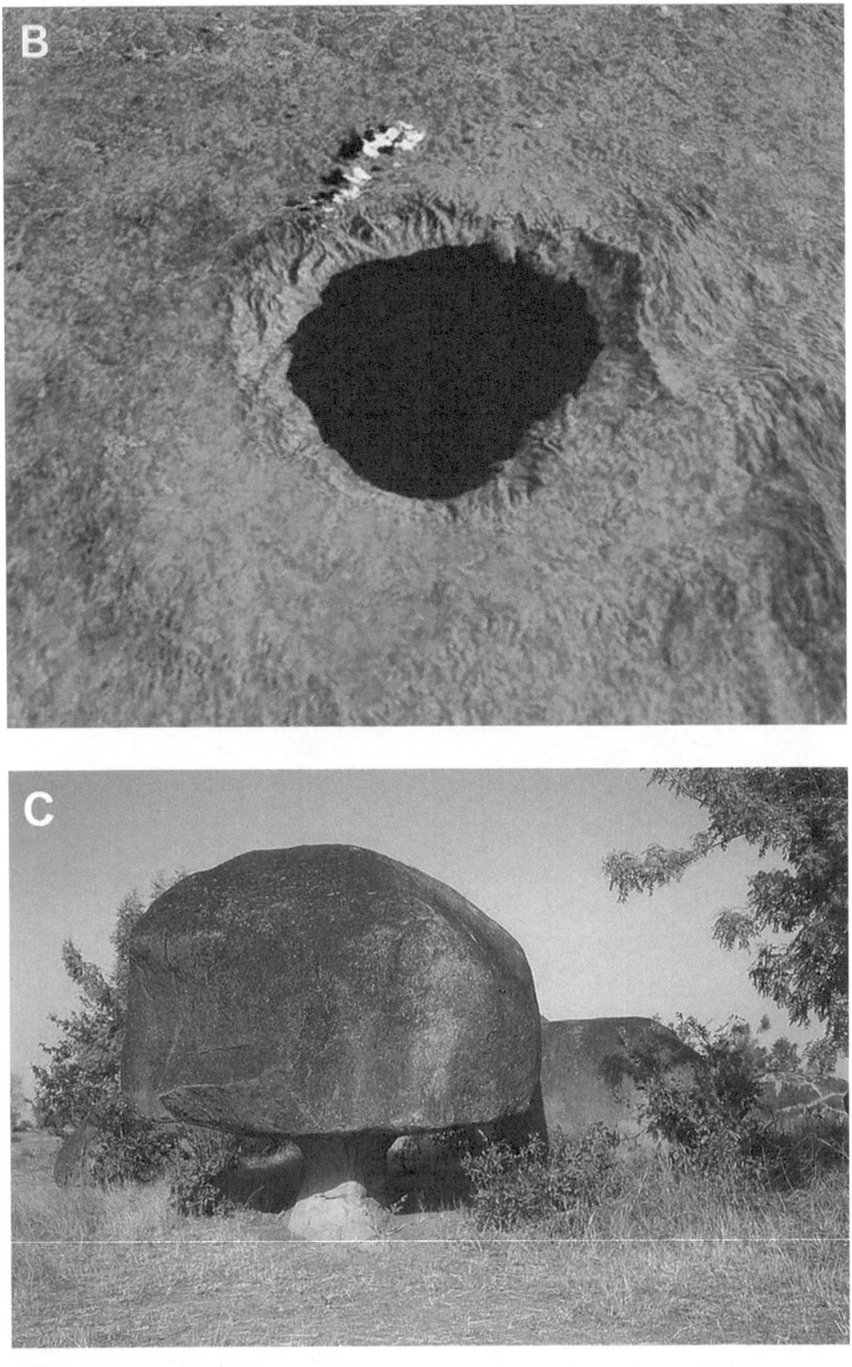

Figure 8.5 (Continued)

Figure 8.5 (Continued)

Delineation of geological sites

Inventory of geological sites involves boundaries, that is, a defined size, scale or magnitude (Brocx & Semeniuk, 2007; de Lima et al., 2010; Brilha, 2016; García-Cortés et al., 2019; Ferrer et al., 2024). Such boundaries can depend on several factors, including geological interest, purpose, land ownership, geomorphology, extent of accessibility, political and administrative boundaries, etc. (Brocx & Semeniuk, 2007). Geosites can be delineated through the use of detailed topographic and geological maps as well as aerial photographs and satellite images, depending on the scale. Boundaries may be identified with morphology (e.g. ridges, watercourses, impact cratering), geological unit or domain (e.g. a mobile belt, sedimentary basin), geodetic points of topographic maps, administrative boundaries (e.g. regions, districts and municipalities), and human-made infrastructure such as roads and railways (Fuertes-Gutiérrez & Fernández-Martínez, 2012; Brilha, 2016; Santos et al., 2020).

The geological sites of interest currently being used for geoscience training by tertiary institutions in Ghana have no boundaries. They range from a few meters to a few kilometres in scale and are located in both rural and urban areas that have extensive human activities. Most of these sites are located in road and railway cuts, coastal communities, and farming areas. Some of the sites are associated with forests and thick vegetation and water bodies.

In delineating these geological sites and other new potential sites of interest for geoscience training by tertiary institutions in Ghana, this study proposes that the delimitation considers mainly the geological interest of the site, since the main purpose is for education and training. Other important factors that should be considered are geomorphology, association with other natural diversity and ecosystem considerations, and extent of accessibility. The sites that are deemed to need it due to their vulnerability and risk of degradation should have the addition of a security perimeter sufficient enough to preserve the geological, physical, and landscape integrity of these sites, without exceeding what is strictly necessary. This is because excessive security measures can lead to unnecessary restriction to access that can hinder public engagement, education, and scientific research. Keeping security measures to the minimum helps balance conservation with accessibility and sustainable use, fostering public support and understanding of geoconservation efforts (Henriques et al., 2011; Brilha, 2016). Detailed geological maps, topographical maps, aerial photographs, and satellite images coupled with ground trothing and local land ownership information should be used to delineate these geological sites.

Assessment of geological sites

The assessment of geological sites for inventory towards geoconservation entails the determination of values, vulnerability, and risk of degradation (Reynard et al., 2007; Fassoulas et al., 2012; Brilha, 2016; Reynard & Brilha, 2017; Brilha, 2018). Geological sites currently being used for geoscience training by tertiary institutions in Ghana have not been assessed for their value. Assessing these sites and other new potential ones should be based on the following:

- The scientific value of a geological site can be defined as its potential to contribute to our understanding of earth's history, processes, and materials, and contribute to geological research (Reynard et al., 2007). The scientific value of these sites should be tied to geological areas of interest such as stratigraphy, geomorphology, palaeontology, tectonics, petrology, geochemistry, metallogeny, mineralogy, hydrogeology, etc. These sites should also be valued for representativeness indicated by the extent to which they exemplify particular geological features, processes, or phenomena, the rarity shown in their uniqueness or the uncommon nature of the geological features present, and their diversity as indicated by the variety of geological features and processes present, all of which provide extensive research opportunities. Also, their state of preservation and completeness of the geological features where well-preserved sites that have not been significantly altered by human or natural processes are particularly valuable for scientific study (Coratza & Giusti, 2005).
- The educational value of a geological site refers to its usefulness in the teaching and learning about geosciences (Mocior & Kruse, 2016; Sanz et al., 2020; Wolniewicz, 2021). This value is assessed by factors such as clarity and accessibility of geological features indicated by well-exposed, easily observable geological formations and processes; representativeness of key geological features and processes that illustrate fundamental concepts in geology, such as

sedimentation, erosion, tectonic activity, and fossilisation; safety and accessibility for field trips indicated by physical accessibility, the presence of infrastructure such as pathways and informational signage, and the overall safety of the site; potential for hands-on learning activities, such as rock sampling, fossil hunting, and geological mapping; and the availability of interpretative resources, such as guides, maps, informational panels, and digital resources. Research has shown that the educational value of a geological site is not always related to its scientific value; for educational purposes, the visibility and comprehensibility of the phenomenon are much more important and dependent on the target group (Drápela, 2022). As such, the assessment of the educational value of the geological sites currently being used for geoscience training by tertiary institutions in Ghana, as well as new potential sites considered for the same purpose, should correlate with their usefulness for visual understanding of geological phenomena as well as for hands-on practice of basic and important geoscience skills such as the identification, description, measurement, analysis, and understanding of geological structures.

- The risk of degradation of a geosite can be defined as the probability of a geological site being damaged or destroyed, and thus losing any of the features that make it valuable as a geological site (García-Ortiz et al., 2014; Herrera-Franco et al., 2023). This can be assessed through the site's fragility and vulnerability to both natural and anthropogenic threats and susceptibility to degradation (García-Ortiz et al., 2014; Selmi et al., 2022). The fragility of a geological site is related mainly to the consolidation and resistance of the lithology. The natural vulnerability has to do with the assessment of the impact or effects of natural processes, both geological and biological, on the site. Vulnerability due to human activities can be assessed by the proximity of the site to human activities, infrastructure, and population. As in the example shown in Figures 8.3c and 8.3d, most of the current geological sites being used for geoscience training by tertiary institutions in Ghana are being damaged or destroyed and are thus losing features that make them valuable as a geological site for geoscience training. There is therefore an urgent need to assess these sites for their risk of degradation and to provide the appropriate conservation solutions.

Detailed guidelines for inventory and assessment with rating scales and numerical evaluations (Wimbledon, 1995; Brilha, 2016; Prosser et al., 2018; García-Cortés et al., 2019) available in the literature can be used to meet the recommended inventory and assessment criteria for the current geological sites being used for geoscience training by tertiary institutions in Ghana and other geological sites that may be considered for the same purpose.

Preservation and protection

The preservation and protection of geological sites involve safeguarding the important geological features of a site (Burek & Prosser, 2008; Gray, 2013). Preservation focuses primarily on maintaining the natural state of geosites while minimising human-induced changes such as construction, pollution, erosion, or vandalism that could degrade or alter the geosites. Protection, on the other hand, encompasses a broader range of activities, including active management and restoration efforts, usually backed by legal designations and regulations (Crofts et al., 2015; Reynard & Brilha, 2017).

The preservation of geosites in Ghana for geoscience training should include techniques such as establishing restricted access areas, implementing regulations to limit human impact, and promoting responsible preservation practices amongst local communities associated with such sites. Protection measures should include restoration efforts to reverse damage already done to the geosites, such as erosion control measures or habitat restoration in areas surrounding the sites. Protection strategies should also involve legal designations, such as national park status or UNESCO Global Geopark designation, which come with specific regulations and management plans aimed at conserving the sites for future generations.

Even though in the laws of Ghana, the Ghana Geological Survey Authority Act, 2016 Act 298 has made provision via Clause 26, Regulation (e) for geoconservation, no sites to date have been identified and developed as geoheritage sites or geosites by law. Geological sites currently being used for geoscience training by tertiary institutions in Ghana, as well as new potential geological sites for the same purpose, should be identified, described, assessed, documented, and proposed for conservation under the Ghana Geological Survey Authority Act, 2016 Act 298. Other institutions such as the EPA, GHC and the Ghana Forestry Commission (GFC) can provide legal designations, policies, and regulations that can be used to protect these sites.

Management

For the effective management of geological sites for geoscience training in Ghana, formal coordination, sources of funding, and a working group responsible for technical and scientific coordination are required (Woo et al., 2013; Migoń & Pijet-Migoń, 2017; Prosser et al., 2018). Institutions in Ghana that should collaborate and take on these responsibilities are MoE, MoLNR, academic institutions, GGSA, GhIG, EPA, GHC, GMMB, and GFC.

The responsibility of formal coordination and funding should be taken up by the MoE and MoLNR, and the technical and scientific coordination should be handled by a workgroup constituted from academic institutions, such as the GGSA, GhIG, EPA, GHC, GMMB, and GFC. All these should work in collaboration with district/municipal and metropolitan assemblies and local communities. Specific management activities should include preparation and implementation of site-specific management plans, the establishment of protocols for accessibility, maintenance and risk management, regular monitoring and maintenance of geological features through physical conservation measures such as erosion control, vegetation management, and restoration of damaged sites to ensure long-term preservation.

The role of geological museums in geoscience training

Museums are a form of ex situ conservation (De Wever & Guiraud, 2018). Generally, geological museums are cultural centres that provide informal, lifelong learning experiences to the general public and they rank as an important instructional approach for the future of geoscience education (Rebbert, 2007; Van Geert, 2022). Generally, the showcasing and highlighting of rock, mineral and fossil collections by geological museums does a great service to our collective understanding of the earth and its evolution and history. The showcased collections also inspire and inform students and young geoscientists in the field of earth science. Even though geological museums traditionally provide informal education to the general public, we believe that a section of the museum could be dedicated to formal geoscience training.

Ghana has only two geological museums: the Department of Earth Science at the University of Ghana museum and the GGSA museum, both of which are open to the general public. At the Department of Earth Science Museum (Figure 8.6), 34 per cent of the museum specimens are minerals, 41 per cent are rocks, and 20 per cent are fossils. The rest of the collections, constituting about 5 per cent, consist of precious stones, models, posters, ores, etc. Of the mineral, rock, and fossil collections, only about 25 per cent are collections from Ghana. The rest of the collections are foreign, mostly from Europe. For the GGSA museum, 10 per cent of the museum specimens are minerals, 74 per cent are rocks, and 5 per cent are fossils. The rest of the collections, constituting about 12 per cent, consist of industrial materials, models, posters, ores, etc. Of the mineral, rock, and fossil collections, about 97 per cent are collections from Ghana. The rest of the collections are foreign, mostly from Germany and Australia.

Figure 8.6 Images from the Department of Earth Science Geological Museum: a) A view of the museum setup; b), c), and d) Displays of rocks, minerals, and fossils.

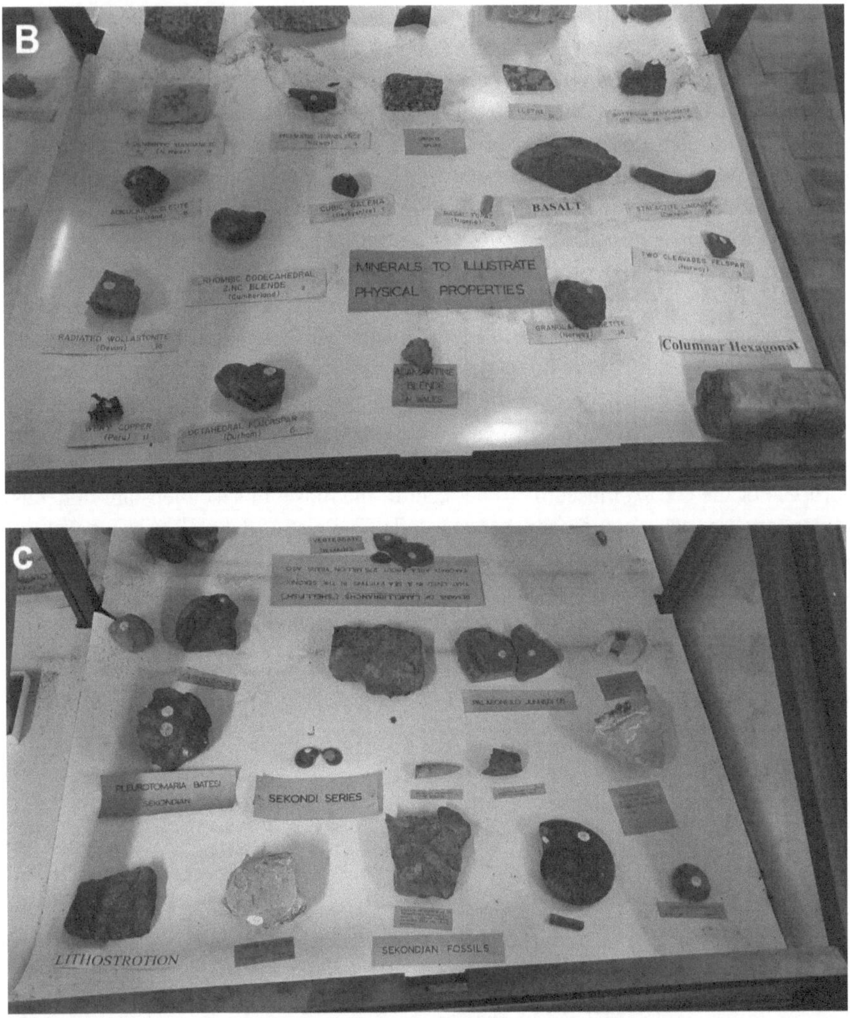

Figure 8.5 (Continued)

The student patronage of the geological museums is generally poor. A reason for the low patronage may be that the collections are not tailored towards their training. To make the museums relevant to geoscience training, this paper recommends that sections of the museums be dedicated to the teaching and learning of the geology of Ghana. All rock units of the geology of Ghana should be captured. Information to be captured may include specimen collection of rock types, geological features, fossils, minerals, and any other important features. Accompanying the specimen collection should be posters displaying the stratigraphy, mineralisation (if any), deformational history, and geological history.

Figure 8.5 (Continued)

Conclusion and recommendations

Ghana is known for its vast geological resources, with gold, manganese, bauxite, and oil and gas being the most prominent. The training of geoscientists is therefore vital for the economic sustainability of the country. A good geoscience training program, however, requires a strong geological field component. Therefore, there is a need to conserve important geological sites for geoscience training. Since geoconservation plays a very crucial role in geoscience training, documentation needs to take place through identification, description, and assessment of geological sites for geoscience training in Ghana, as well as preserving, protecting, and appropriately managing such sites.

It is recommended that the geoscience community in Ghana (including academic institutions offering geoscience programs in Ghana, the Ghana Geological Survey Authority, geoscience professionals who carry out fieldwork, and the Ghana Institute of Geoscientists) collaboratively take on the responsibility of planning, organising, facilitating, and carrying out field missions for the identification, description, and assessment of geosites for geoscience training, and these identified groups in turn collaborate and partner with state agencies such as the district/municipal and metropolitan assemblies, EPA, MoE, MoLNR, GHC, GFC, and the GMMB, as well as local communities and private interests, in the management and protection of these geosites. It is also recommended that geological museums be adequately equipped and organised for use for formal geoscience training.

References

Adade Williams, P., Sikutshwa, L., & Shackleton, S. (2020). Acknowledging indigenous and local knowledge to facilitate collaboration in landscape approaches-Lessons from a systematic review. *Land*, *9*(9), 331.

Agyei Duodu, J., Loh, G.K., Boamah, K.O., Baba, M., Hirdes, W., Toloczyki, M., & Davis, D.W. (2009). *Geological map of Ghana 1:1 000 000*. Geological Survey Department of Ghana (GSD).

Arukwe, C.M., Ezenwa, E.E., Meniru, I.C., & Emenaha, O.T. (2022). Strong industry-academia relationship: A tool to improve the quality of geoscience education in Nigerian public universities. *Journal of Educational Studies*, *4*(1), 81–107.

Baoguo, C.H.E.N., Jiuchen, Z.H.A.N.G., & Mengmeng, Y.A.N.G. (2016). The present research and prospect of Chinese geosciences history. *Acta Geologica Sinica-English Edition*, *90*(4), 1516–1534.

Brilha, J. (2016). Inventory and quantitative assessment of geosites and geodiversity sites: A review. *Geoheritage*, *8*(2), 119–134.

Brilha, J. (2018). Geoheritage: Inventories and evaluation. In E. Reynard & J. Brilha (Eds.), *Geoheritage: Assessment, protection and management*, pp. 69–85. Amsterdam, Netherlands: Elsevier.

Brocx, M., & Semeniuk, V. (2007). Geoheritage and geoconservation-history, definition, scope and scale. *Journal of the Royal Society of Western Australia*, *90*(2), 53–87.

Brown, E.J., Evans, D.H., Larwood, J.G., Prosser, C.D., & Townley, H.C. (2018). Geoconservation and geoscience in England: A mutually beneficial relationship. *Proceedings of the Geologists' Association*, *129*(3), 492–504.

Burek, C.V., & Prosser, C.D. (2008). The history of geoconservation: An introduction. *London: The Geological Society of London*, *300*(1), 1–5.

Butler, R. (2008). *Teaching geoscience through fieldwork*. Plymouth, England: University of Plymouth.

Capello, M.A., & Howes, C.S. (2022, August). Attracting students to geoscience careers in energy: Analysis of opportunities and challenges. In *Second international meeting for applied geoscience & energy*. Society of Exploration Geophysicists and American Association of Petroleum Geologists, pp. 3311–3315.

Coratza, P., & Giusti, C. (2005). Methodological proposal for the assessment of the scientific quality of geomorphosites. *Alpine and Mediterranean Quaternary*, *18*(1), 307–313.

Crofts, R., Gordon, J.E., Brilha, J.B., Gray, M., Gunn, J., Larwood, J., Santucci, V.L., Tormey, D., & Worboys, G.L. (2020). *Guidelines for geoconservation in protected and conserved areas*. Gland, Switzerland: IUCN.

Crofts, R., Gordon, J.E., & Santucci, V.L. (2015). Geoconservation in protected areas. *Protected area governance and management*. Canberra: ANU Press, pp. 531–568.

de Lima, F.F., Brilha, J.B., & Salamuni, E. (2010). Inventorying geological heritage in large territories: A methodological proposal applied to Brazil. *Geoheritage*, *2*, 91–99.

De Wever, P., & Guiraud, M. (2018). Geoheritage and museums. In E. Reynard & J. Brilha (Eds.), *Geoheritage: Assessment, protection and management*, pp. 129–145. Amsterdam, Netherlands: Elsevier.

Drápela, E. (2022). Assessing the educational potential of geosites: Introducing a method using inquiry-based learning. *Resources*, 11(11), 101.

Elkins, J.T., & Elkins, N.M. (2007). Teaching geology in the field: Significant geoscience concept gains in entirely field-based introductory geology courses. *Journal of Geoscience Education*, *55*(2), 126–132.

Fassoulas, C., Mouriki, D., Dimitriou-Nikolakis, P., & Iliopoulos, G. (2012). Quantitative assessment of geotopes as an effective tool for geoheritage management. *Geoheritage*, *4*, 177–193.

Ferrer, N., Vegas, J., & Díez-Herrero, A. (2024). The magnitude rule in geoheritage site inventories. *Progress in Physical Geography: Earth and Environment*, *48*(2), 191–207.

Fuertes-Gutiérrez, I., & Fernández-Martínez, E. (2012). Mapping geosites for geoheritage management: A methodological proposal for the Regional Park of Picos de Europa (León, Spain). *Environmental Management*, *50*, 789–806.

García-Cortés, A., Vegas, J., Carcavilla, L., & Díaz-Martínez, E. (2019). *Conceptual base and methodology of the Spanish inventory of sites of geological interest (IELIG)*. Madrid: Instituto Geológico y Minero de España, p. 102.

García-Ortiz, E., Fuertes-Gutiérrez, I., & Fernández-Martínez, E. (2014). Concepts and terminology for the risk of degradation of geological heritage sites: Fragility and natural vulnerability, a case study. *Proceedings of the Geologists' Association, 125*(4), 463–479.

Gray, M. (2013). *Geodiversity: Valuing and conserving abiotic nature*. Chichester, England: Wiley-Blackwell.

Henriques, M.H., dos Reis, R.P., Brilha, J., & Mota, T. (2011). Geoconservation as an emerging geoscience. *Geoheritage, 3*, 117–128.

Herrera-Franco, G., Mata-Perelló, J., Caicedo-Potosí, J., & Carrión-Mero, P. (2023). Vulnerability in geosites: A systematic literature review. *Sustainability in Practice: Addressing Challenges and Creating Opportunities in Latin America*, 395–407.

IUGS-International Commission on Geoheritage. (2023). *Guidelines for the assessment of the international significance of geological heritage in UNESCO global geopark applications*. Paris, France: IUGS.

Kastens, K.A., & Manduca, C.A. (Eds.). (2012a). *Earth and mind II: A synthesis of research on thinking and learning in the geosciences*. Colorado, United States of America: Geological Society of America.

Kastens, K.A., & Manduca, C.A. (2012b). *Mapping the domain of field-based teaching and learning in the geosciences. Earth and Mind II: A synthesis of research on thinking and learning in the geosciences*. Colorado, United States of America: Geological Society of America, pp. 125–130.

Kesse, G.O. (1985). *The mineral and rock resources of Ghana*. Rotterdam: Balkema. ISBN: 9789061915973

King, C., Gorfinkiel, D., & Frick, M. (2021). International comparisons of school-level geoscience education – the UNESCO/IGEO expert opinion survey. *International Journal of Science Education, 43*(1), 56–78.

LaDue, N.D., & Pacheco, H.A. (2013). Critical experiences for field geologists: Emergent themes in interest development. *Journal of Geoscience Education, 61*(4), 428–436.

Martínez-Frías, J., & Mogessie, A. (2012). The need for a geoscience education roadmap for Africa. *Episodes Journal of International Geoscience, 35*(4), 489–492.

Migoń, P., & Pijet-Migoń, E. (2017). Viewpoint geosites – values, conservation and management issues. *Proceedings of the Geologists' Association, 128*(4), 511–522.

Mocior, E., & Kruse, M. (2016). Educational values and services of ecosystems and landscapes – an overview. *Ecological indicators, 60*, 137–151.

Mogk, D.W., & Goodwin, C. (2012). *Learning in the field: Synthesis of research on thinking and learning in the geosciences*. Colorado, United States of America: Geological Society of America, pp. 131–163.

North, M.A., Hastie, W.W., & Hoyer, L. (2020). Out of Africa: The underrepresentation of African authors in high-impact geoscience literature. *Earth-Science Reviews, 208*, 103262.

Prosser, C.D., Brown, E.J., Larwood, J.G., & Bridgland, D.R. (2013). Geoconservation for science and society – an agenda for the future. *Proceedings of the Geologists' Association, 124*(4), 561–567.

Prosser, C.D., Díaz-Martínez, E., & Larwood, J.G. (2018). The conservation of geosites: Principles and practice. In *Geoheritage*. Elsevier, pp. 193–212.

Rebbert, C.R. (2007). The role of museums in geoscience education: A perspective. *The Leading Edge, 26*(10), 1322–1325.

Reynard, E., & Brilha, J. (Eds.). (2017). *Geoheritage: Assessment, protection, and management*, p. 450. Amsterdam, Netherlands: Elsevier.

Reynard, E., Fontana, G., Kozlik, L., & Scapozza, C. (2007). A method for assessing "scientific" and "additional values" of geomorphosites. *Geographica Helvetica, 62*(3), 148–158.

Santos, D.S., Mansur, K.L., Seoane, J.C., Mucivuna, V.C., & Reynard, E. (2020). Methodological proposal for the inventory and assessment of geomorphosites: An integrated approach focused on territorial management and geoconservation. *Environmental Management, 66*, 476–497.

Sanz, J., Zamalloa, T., Maguregi, G., Fernandez, L., & Echevarria, I. (2020). Educational potential assessment of geodiversity sites: A proposal and a case study in the Basque Country (Spain). *Geoheritage*, *12*, 1–13.

Sapah, M.S., Asiedu, D.K., & Loh, Y.S.A. (2024). Geoscience education in Ghana: Status and recommendations. *Episodes Journal of International Geoscience*, *47*(1), 1–9.

Selmi, L., Canesin, T.S., Gauci, R., Pereira, P., & Coratza, P. (2022). Degradation risk assessment: Understanding the impacts of climate change on geoheritage. *Sustainability*, *14*(7), 4262.

Sharples, C. (1993). *A methodology for the identification of significant landforms and geological sites for geoconservation purposes.* Hobart, Australia: Forestry Tasmania.

Sharples, C. (2002). *Concepts and principles of geoconservation.* Hobart, Australia: Tasmanian Parks & Wildlife Service.

Sorkhabi, R. (2023). Geoscience education and workforce development for energy transition. *Current Science*, *125*(11), 1173.

Stokes, A., Boyle, A.P., & Whitmeyer, S.J. (2009). The undergraduate geoscience fieldwork experience: Influencing factors and implications for learning. *Field Geology Education: Historical Perspectives and Modern Approaches*, *461*, 291.

Tavares, A.O., Henriques, M.H., Domingos, A., & Bala, A. (2015). Community involvement in geoconservation: A conceptual approach based on the geoheritage of South Angola. *Sustainability*, *7*(5), 4893–4918.

Van Geert, F. (2022). The exhibition of geoheritage in geoparks and geological museums. *Museologia e Patrimônio*, *15*(1).

Whitmeyer, S.J., & Mogk, D.W. (2009). Geoscience field education: A recent resurgence. *Eos, Transactions American Geophysical Union*, *90*(43), 385–386.

Wimbledon, W.A. (1995). The development of a methodology for the selection of British geological sites for conservation: Part 1. *Modern Geology*, *20*, 159–202.

Wolniewicz, P. (2021). Beyond geodiversity sites: Exploring the educational potential of widespread geological features (rocks, minerals and fossils). *Geoheritage*, *13*(2), 34.

Woo, K., Sohn, Y., Ahn, U., Yoon, S., & Spate, A. (2013). Management plan. In *Jeju Island geopark – A volcanic wonder of Korea. Geoparks of the world (closed)*, Vol. 1. Berlin, Heidelberg: Springer. https://doi.org/10.1007/978-3-642-20564-4_10

9 Conclusion

Advancing geotourism for social sustainability and ecological justice in sub-Saharan Africa

Llewellyn Leonard and Khodani Matshusa

Introduction

The journey through the geological wonders and potential of sub-Saharan Africa (SSA) has illuminated a crucial path forward for the region's development. The narratives, case studies, and analyses presented in this book have underscored the immense value of SSA's geological heritage, not only as a repository of natural beauty but also as a catalyst for sustainable socioeconomic development, environmental conservation, and social cohesion. This conclusion synthesizes the key insights gleaned from the preceding chapters and proposes actionable steps to harness the full potential of geotourism in advancing social sustainability and ecological justice in SSA. As this book has strongly highlighted SSA geographic landscape is endowed with a rich geological heritage that remains underexplored and underappreciated. Despite this abundance, geotourism in SSA has not received the same level of attention and development as in other parts of the developed world. As was highlighted in Chapter 1, scholars such as Matshusa and Leonard (2023) and Quesada-Valverde and Quesada-Román (2023) have highlighted this academic gap, underscoring the need for greater exploration and promotion of Africa's geological wonders. Most geotourism research is concentrated in regions like Europe, Asia, and South American and country leaders such as China, Iran, and Brazil, receiving considerable attention (Ruban, 2015). Africa and SSA are lagging in both scholarly inquiry and practical initiatives. Geotourism in SSA presents a unique opportunity to bridge this gap by fostering a deeper connection between people and their natural environment.

Geotourism promotes environmental sustainability by encouraging the conservation of geological sites and landscapes, which in turn supports biodiversity and ecological health (Ólafsdóttir and Tverijonaite, 2018). Economically, geotourism can serve as a powerful engine for local development, providing employment opportunities, stimulating local businesses, and attracting investment (Matshusa et al., 2021). Socially, it has the potential to foster a sense of pride and ownership among local communities, empowering them to engage actively in the stewardship of their natural and cultural heritage (Newsome and Dowling, 2018). Thus, geotourism development holds the potential to yield significant economic benefits for local communities (Sagala et al., 2018; Gupta et al., 2023). This book presents the

DOI: 10.4324/9781032706764-9

reality that one need not be oblivious to the reality that the advancement of geotourism in SSA faces numerous challenges. High unemployment rates, pervasive poverty, and a lack of infrastructure can hinder the development and sustainability of geotourism initiatives, including threats such as climate change and industrial activities. The interplay between economic interests, environmental preservation, and community welfare is complex and often contentious, necessitating a balanced and inclusive approach to geotourism development. Each of the chapters in this book presented geotourism potential for development in different context.

In Chapter 2, Donkor, Tantoh, and Mearns explored the role of geotourism in fostering local economic development in South Africa, examining both the constraints and opportunities associated with this niche tourism sector. With limited research conducted on geotourism's impact on local development, the chapter aimed to investigate its potential to drive grassroots economic growth. The chapter highlighted a significant issue: low awareness of geotourism in South Africa, which has hindered its development. This lack of awareness has resulted in inadequate support for tourists interested in visiting geoheritage sites, ultimately affecting local economic benefits. Donkor et al. emphasised the critical role of effective packaging in tourism, noting that it enhances diversification and specialisation. Poor packaging and marketing of geotourism products can diminish the appeal of a destination. Infrastructure is also crucial for facilitating visits to geotourism sites. This includes providing ecologically friendly amenities such as roads and toilets near geoheritage sites, with consideration for visitors with disabilities. Additionally, safety and security are non-negotiable factors essential for sustainable tourism growth. The chapter also stressed the importance of preserving the sacredness and spirituality of geotourism sites in alignment with cultural heritage values. Adequate legal and policy support is necessary to ensure industry standards are upheld, protecting both customer needs and the environment. Local businesses require support to thrive and become pioneers in this emerging tourism sector. Addressing everyday issues of safety, security, and ethics is vital for creating a conducive operating environment that enables the geotourism sector to remain competitive.

Chapter 3 by Dyssel conducts a social-ecological analysis of geoheritage resource uses and livelihoods concerns of villagers in the Cederberg Mountains, Western Cape, South Africa. The Cederberg Mountains in the Western Cape province of South Africa are one of the true wilderness areas in the country. This mountainous region is rich in biodiversity, geodiversity, and cultural-historical diversity. Dyssel analysed, through the theoretical lens of Social-Ecological Systems, the perceptions and concerns of villagers regarding sustainable access to their geoheritage resources to sustain their livelihoods. However, residents of remote hamlets like Wupperthal, Heuningvlei, Eselbank, and Langkloof were largely excluded from the tourism and livelihood opportunities available in the area. The ancient presence of the First Peoples, including the San (hunter-gatherers) and Khoikhoi (pastoralists), was evident at various rock art sites in the Cedarberg. These sites depicted a different historical ecology with well-preserved sketches of eland, elephants, and leopards that once roamed the area. The Cedarberg's geology and geomorphology featured spectacular landscapes and landforms, predominantly composed of

Table Mountain Sandstone, visible in high mountain peaks like Sneeuberg, Sneeukop, and Tafelberg. Iconic landforms such as the Wolfberg Arch, Stadsaal Caves, and Maltese Cross were major attractions. Participants in all villages were unanimous in their appreciation of any new initiatives that would contribute to their livelihoods by promoting and developing their own geoheritage and cultural resources in sustainable ways. The uniqueness of the study, therefore, lies in the opinions of locals on how to sustainably benefit from commercialised and non-commercialised geoheritage sites in and around an ecologically-sensitive wilderness area. Dyssel notes that the spatial footprint of South Africa's conservation estate needs to be extended based on its rich conservation endowment, however, it needs to be done with long-term sustainable participatory synergies and action plans with local people for future success of the CWA and its surroundings.

In Chapter 4, Matshusa examined the impacts of climate change on geoheritage and geotourism in Southern Africa. This study sought to locate and understand the impacts of climate change on selected geoheritage sites in Southern Africa that might have become obscured. By focusing on the impacts of climate change in Southern Africa and chapter explored three case sites: the Cradle of Humankind World Heritage Site, the Magaliesberg Biosphere Reserve and the Kruger National Park. Matshusa highlighted that geoheritage sites are prone to impacts of climate change, such as high temperatures, heavy rainfall, and floods. The study revealed impacts of climate change on geoheritage in the Global South that are also relevant and applicable to the Global North. Extreme climatic conditions and events caused environmental damage to geoheritage sites in the study areas through weathering, erosion, flooding, pollution, and tree and plant encroachment. These have the potential to damage the scientific value, aesthetic value, and integrity of geoheritage sites and geotourism. Flooding has also resulted in damage and collapse of infrastructure for geoheritage and geotourism. In this context, the chapter notes that legislation, policies, strategies, and environmental plans on climate change adaptation must support tourism in view of extreme climate events such as droughts, fires, and floods. This legislation and strategies on climate change must take the geoheritage sites in Southern African into account.

In Chapter 5, Leonard conducted a critical examination of the political economy of mining development impacts for sustainable development and geotourism. The mining industry has been a dominant force in the SSA economy for decades, with significant impacts on conservation, the environment, and communities. This has the potential to undermine the region's geotourism potential. The chapter shed light on the complex interplay between mining development and geotourism in Southern Africa. The undeniable economic significance of the mining industry poses challenges to the preservation of geological wonders that are essential for geotourism. This is considering that Southern Africa is rich in mineral wealth and its countries will continue to harness their geological endowments, particularly in gold, diamonds, platinum, and other precious minerals, driving economic growth and attracting foreign investment. The exploration of the impact of political economy industrial processes, specifically mining development, on geotourism sites and local economic development remains relatively uncharted territory in academic discourse. Leonard notes how local economies and communities, often intertwined with (geo)tourism, stand to experience losses in livelihoods and long-term employment opportunities due to mining destruction of sites with geotourism potential. SSA has also grappled with the detrimental consequences of poor governance and widespread corruption, particularly in the context of mining development, having implications for geotourism and its development in the region. The author highlights that addressing governance challenges, including mining and geotourism challenges, necessitates a nuanced approach, emphasising stakeholder collaboration with local communities and effective policies implemented to protect geotourism interests for the benefit of the region. This chapter underscored the imperative of a critical assessment of the nexus between mining development and geotourism to foster sustainable tourism and achieve overarching sustainable development goals.

In Chapter 6, Tantoh, Donkor, and McKay examined the potential contribution of geotourism and environmental protection for sustainable development in Cameroon. The country is renowned for its outstanding and abundant natural resources and erratic natural beauty, with numerous remarkable geological features. Despite the remarkable geoheritage sites studies on geoconservation were noted to be limited. This chapter investigated geoheritage sites in Cameroon and their contribution to geoconservation. By promoting responsible tourism practices, fostering a sense of environmental stewardship and offering economic benefits to local communities, geotourism can help to strike a balance between preserving the country's ecological treasures and fostering economic growth. As the country strives for a sustainable future, geotourism offers a path forward that respects the environment while embracing the potential of responsible tourism. It presents an opportunity to address environmental challenges by promoting responsible tourism practices that respect the environment. The authors note that collaboration among governments, local communities, tourists, and environmental organisations is essential for achieving this delicate balance between geotourism and environmental protection. In addition, the integration of geotourism and environmental protection can offer a promising path towards sustainable development in Cameroon. By leveraging the country's geological heritage, Cameroon can enhance its tourism sector, support local communities, and protect its unique natural environment for future generations.

In Chapter 7, Tefogoum, Dedzo, Dama, Ndonbou, Dongmo, and Nkouathio explored a geological heritage for geotouristic and geoeducational practices at the Mount Bambouto slopes geomorphosites in Cameroon. It focused on the slopes of the massif to carry out an inventory and study of the geomorphosites. The chapter highlighted the scientific and aesthetic value of the geomorphosites of the latter. The results of this study also completed the database of the Mount Bambouto geomorphosites, which can be used by decision-makers for the implementation of sustainable geotourism and geoconservation strategies in that locality. The Mount Bambouto slopes are rich in geomorphosites and characterised by several diverse geomorphosites, mostly of volcanic origin. These include domes, dykes, the Totap cone, escarpments and waterfalls, potholes, and depression lakes. Some of these sites are passive, while others are active. However, their cultural, geoeducational, and geotourism value is quite significant. The chapter recommends that these geomorphosites be promoted and preserved to foster recreation activities and long-term field training of students from national institutions. Tefogoum's study of the Mount Bambouto slopes geomorphosites underscores their immense potential for contributing to sustainable geotourism and geoeducation in Cameroon. By promoting and preserving these geological treasures, stakeholders can enhance the region's appeal as a geotourism destination while providing valuable educational opportunities and supporting local communities. The chapter serves as a call to action for comprehensive strategies that balance economic development with ecological and cultural preservation.

In Chapter 8, Sapah, Asiedu, Loh, and Banoeng-Yakubo focussed on geoscience training in Ghana's through geoconservation strategies. This chapter discussed the need for geoconservation as a means of advancing geoscience training in tertiary institutions in Ghana, by identifying, documenting, and managing important geological sites. The chapter argued that Ghana has no known geoheritage sites dedicated to geoscience education and training. The study highlighted that geoheritage sites in Ghana are being destroyed mostly due to human activities since there are no conservation laws to protect them. The authors recommend methods of geoconservation for geoscience training in Ghana, including inventory, preservation and protection, and management. This chapter identified some geological sites of interest, as recommendations to the Ghana Geological Survey Authority for geoconservation by law. The authors advocate for the involvement of various stakeholders such as academic institutions in Ghana, the Ghana Geological Survey Authority, geoscience professionals, the Ghana Institute of Geoscientists, state agencies, local communities, and private interests to collaborate in planning, organising, facilitating, and carrying out field missions for the identification, description and assessment of geosites for geoscience training in Ghana. This will lead to their preservation as natural classrooms for geoeducation and training in Ghana.

Advancing geotourism in sub-Saharan Africa: pathways to sustainable development

The potential for geotourism to drive sustainable development in SSA is immense, given the region's rich and diverse geological heritage. As the chapters in this book have highlighted, geotourism offers a unique avenue to balance economic growth, environmental conservation, and social sustainability. By showcasing geological wonders, fostering community engagement, and promoting environmental stewardship, geotourism can play a pivotal role in sustainable development across the continent. As highlighted, geotourism's importance transcends economic benefits, acting as a catalyst for innovative solutions that enable the integration of local communities with their geoheritage resources and cultural legacies (Matshusa et al., 2021). Geotourism offers a unique platform for promoting both conservation efforts and socio-economic development by fostering a harmonious relationship between local communities, broader society, and wildlife. Additionally, it holds promise as a proactive response to the threats facing geoheritage sites, including the impacts of climate change and industrial development in pristine environments. However, as most chapters have noted, the development of geotourism in SSA faces significant challenges, including limited and fragmented research, inadequate infrastructure, and a lack of comprehensive policies and governance frameworks. While countries like South Africa and Tanzania have made strides in promoting geotourism, much of the continent remains untapped in this regard. Considering the academic gap in this area, requires concerted effort to elevate geotourism in the region.

One avenue for advancing geotourism is through the African Union (AU), which can play a pivotal role in advancing geotourism in SSA as a continental organisation to coordinate, support, and promote initiatives that harness the region's geological heritage for sustainable development. Considering that the AU strives to promote Africa's growth and economic development by championing citizen inclusion and increased cooperation and integration of African states, this can serve as a platform to raise awareness and promote geotourism development. As noted by the AU (2014), the African Union Ministerial Working Group on tourism sector development strategy have been working to strategize on how to make an African brand through tourism while attracting visitors to choose Africa as major destination. It was noted that Africa has a huge tourism potential in all aspects of the sector, such as natural environmental wonders, exotic wildlife and plants, ancient cultures, and historic relics. Regional organisations such as the Geological Society of Africa (GSAf) and Organisations of African Geological Surveys (OAGS) are also charged with broader promotion of geoscience activities in Africa. The International Union of Geological Sciences (IUGS) Ten Action Plan for Africa "TAP for Africa" (IUGS, 2024) seeks to partner with such organisations to implement and support these actions. Action 10 of the "TAP for Africa" was established to:

- Promote research training and specific tools for a better geological knowledge specifically on Earth's resources.
- Increase the awareness of geosciences, geoconservation and the potential of Geotourism in Africa. Include more African Geological Heritage Sites and Heritage Stones in the IUGS designation programs and explore the potential to link to UNESCO Global Geoparks as tools for local development.

The African Geoparks Network which was established in 2019 with the aim of enhancing collaboration within the designated geoparks and highlighting Africa's rich geoheritage and culture while ensuring local community development. All these have not yet been well exploited and promoted to the extent of generating commensurate benefit to African countries. Therefore, geoheritage and geotourism needs to become part of the ministerial agenda for sustainable development. Thus, the following recommendations are made for the AU and governments to enable a geotourism agenda on the continent:

Policy formulation, harmonisation and good governance: There needs to be the development of a comprehensive continental policy framework for geoheritage and geotourism that aligns with the AU Agenda 2063, which aims for inclusive and sustainable development. This policy should include guidelines for geoheritage conservation, sustainable tourism practices, and community engagement. By harmonising geotourism policies across member states can ensure consistent standards and practices. Furthermore, integrating geoconservation strategies into broader environmental and biodiversity conservation programs can work to protect geoheritage sites from degradation and unsustainable exploitation such as from unsustainable industrial development. Governments should enact and enforce regulations that protect geoheritage sites, promote sustainable tourism practices, and ensure that the benefits of geotourism are equitably distributed. Collaborative governance models that involve multiple stakeholders, including local communities, private sector partners, and civil society organisations, can enhance transparency, accountability, and inclusivity. Therefore, there is a need for African countries to immediately establish an effective multinational cooperation mechanism to unify efforts on geoheritage and geotourism development in Africa.

Learning from best practice: There must be the creation of a set of best practices and standards for geoheritage and geotourism development that African member states can adopt and adapt or use as guidelines for inventory and geoconservation of geoheritage sites. This must involve working with key academic institutions that work on geoheritage, geotourism, geoparks, and to document best practices and challenges. For example, the University of South Africa, Centre for Excellence: Adaptation and Resilience is one of the research hubs that explore geoheritage and geotourism in South Africa and expanding across the SSA region through networks and partnerships. As such the AU must work with leading organisations such as the United Nations Educational, Scientific and Cultural Organization (UNESCO), IUGS, and International Union for Conservation of Nature (IUCN) to create a central database or repository for geoheritage sites and geotourism resources in SSA, which can be accessible to researchers, policymakers, and tourism operators. Academic institutions, government bodies, and international organisations must collaborate to conduct comprehensive studies, identify key geosites, and assess their potential for geotourism. This knowledge base will be crucial for informed decision-making, strategic planning, and geoconservation.

Capacity building, training and collaboration: There must be the implementation of capacity-building programs to train local communities, government officials, academics, and tourism operators on geoheritage and sustainable geotourism practices. This can include workshops, seminars, and exchange programs with regions that have successfully developed geotourism. As such educational curricula and materials focused on geoheritage and geotourism for integration into schools and universities across the continent can begin to promote awareness about its development capacity. In this sense, the editors suggest that geoeducation can be used as tool for enhancing geoconservation, while creating geoscience awareness for the students and local communities where these geoheritage sites are located. International frameworks like the UNESCO Global Geoparks Network and International Commission on Geoheritage (ICG) of the IUGS have been created to promote the importance and significance of protecting the Earth's geodiversity through actively engaging with the local communities and enhancing collaboration and partnerships with other organisations at global, regional, and local levels. These resulted in the designation of 213 UNESCO Global Geoparks with only two in Africa as of August 2024. The ICG has thus far left much to be desired in terms of fully and actively collaborating with African networks and researchers and putting African geoheritage at the centre of its agenda and projects. Although the African Geoparks Network was established in 2019, its impact and effectiveness are far from being desirable to non-existence. Therefore, the newly established Union of African Geoheritage and Geotourism (UAGG) will play a pivotal role to enhance capacity building and harmonize the efforts on geoheritage, geotourism, and geoparks in Africa. The AUGG will also provide a unique platform for international organisations like IUGS, ICG, IUCN, researchers, and other key stakeholders such as local communities to collaborate in research and development on African geoheritage and geotourism. By fostering collaboration between researchers, governments, private sector, NGOs, and international organisations to support geotourism initiatives can also ensure multi-stakeholder platforms for dialogue and cooperation. Such platforms can also encourage and support community-based geotourism initiatives that involve local populations in the management and promotion of geoheritage sites so that the economic benefits of geotourism are equitably distributed. Empowering local communities is at the heart of successful geotourism initiatives. This will involve the facilitation and formation of community cooperatives or associations that can manage geotourism activities and advocate for local interests.

Infrastructure development: There must be support for the development of infrastructure that facilitates geotourism, such as roads, tour trails, visitor centres, and accommodation near key geoheritage sites. Improving infrastructure and accessibility to geoheritage sites is essential for attracting tourists and facilitating local engagement. Investments in transportation, tour trails, visitor facilities, and interpretive centres or panels can enhance the overall geotourism experience while ensuring the sustainability and conservation of these sites. This can include partnerships with private sector investors and development agencies. It is encouraging that the AU has as its vision to enable tourism infrastructure on the continent, which must be enabled as priority. At the same time there must be the need to promote the use of sustainable and environmentally friendly infrastructure to minimize the ecological footprint of such tourism activities for protection of nay geotourism sites.

These recommendations are in harmony with the IUGS Ten Action Plan for Africa "TAP for Africa", especially action 4 to "Improve Geological Knowledge and Increase the Awareness of Geological Sciences" (IUGS, 2024). By taking these steps, the African Union and governments can significantly advance geotourism in the region, creating sustainable economic opportunities, preserving geological heritage, and enhancing the well-being of local communities. As this book has repeatedly emphasised, advancing geoheritage and geotourism in SSA holds transformative potential for the region's sustainable development. By embracing the geological richness of SSA and fostering a holistic approach that integrates environmental conservation, social equity, and economic development, geotourism can serve as a powerful tool for social sustainability and ecological justice. The insights and recommendations presented in this book aim to inspire stakeholders across various sectors to collaborate, innovate, and take decisive action towards realising the full potential of geotourism in SSA. Through collective effort and shared vision, it is possible to pave the way for a more inclusive, sustainable, and prosperous future, where the geological heritage of SSA is celebrated, conserved, and leveraged for the benefit of all.

References

African Union. (2014). *Meeting of the Ministerial Working Group on the long-term Tourism Development Strategy in Africa (African Union Agenda 2063)*, 13 March. https://au.int/en/newsevents/20140313/meeting-ministerial-working-group-long-term-tourism-development-strategy-africa (Accessed on 19 May 2024).

Gupta, A., Zhu, H., Bhammar, H., Earley, E., Filipski, M., Narain, U., Spencer, P., Whitney, E. and Taylor, J.E. (2023). Economic Impact of Nature-Based Tourism. *PLoS ONE*, 18(4), pp. 1–16. https://doi.org/10.1371/journal.pone.0282912.

International Union of Geological Sciences. (2024). *IUGS Geoscience in Africa Event: IUGS Ten Action Plan for Africa "TAP for Africa"*. www.iugs.org/_files/ugd/f1fc07_f37dad4f-2b4049f89233a0a3224a5c5d.pdf?index=true (Accessed on 20 May 2024).

Matshusa, K. and Leonard, L. (2023). Proposed Governance Model for Geotourism and Geoheritage in South Africa and the Implications for Africa. *Geoheritage*, 15, pp. 1–18. https://doi.org/10.1007/s12371-022-00783-9.

Matshusa, K., Leonard, L. and Thomas, P. (2021). The Contribution of Geotourism to Social Sustainability: Missed Opportunity? *The International Journal of Social Sustainability in Economic, Social, and Cultural Context*, 17, pp. 95–118. https://doi.org/10.18848/2325-1115/CGP/v17i01/95-118.

Newsome, D. and Dowling, R. (2018). Geoheritage and Geotourism. In: E. Reynard and J. Brilha (Eds.), *Geoheritage – Assessment, Protection and Management*. Amsterdam: Elsevier. https://doi.org/10.1016/B978-0-12-809531-7.00017-4.

Ólafsdóttir, R. and Tverijonaite, E. (2018). Geotourism: A Systematic Literature Review. *Geosciences*, 8(7), pp. 1–16. https://doi.org/10.3390/geosciences8070234.

Quesada-Valverde, M.E. and Quesada-Román, A. (2023). Worldwide Trends in Methods and Resources Promoting Geoconservation, Geotourism, and Geoheritage. *Geosciences*, 13(2), pp. 1–12. https://doi.org/10.3390/geosciences13020039.

Ruban, D.A. (2015). Geotourism – A Geographical Review of the Literature. *Tourism Management Perspectives*, 15, pp. 1–15. https://doi.org/10.1016/j.tmp.2015.03.005.

Sagala, S., Rosyidie, A., Sasongko, M.A. and Syahbid, M.M. (2018). Who Gets the Benefits of Geopark Establishment? A study of Batur Geopark Area, Bali Province, Indonesia. *IOP Conference Series: Earth and Environmental Science*, 158, pp. 1–16. https://doi.org/10.1088/1755-1315/158/1/012034.

Index

Note: Page numbers in *italics* indicate figures and page numbers in **bold** indicate tables.